MOOC Courses and the Future of Higher Education

A New Pedagogical Framework

RIVER PUBLISHERS SERIES IN INNOVATION AND CHANGE IN EDUCATION - CROSS-CULTURAL PERSPECTIVE

Indexing: All books published in this series are submitted to the Web of Science Book Citation Index (BkCI), to CrossRef and to Google Scholar for evaluation and indexing.

Nowadays, educational institutions are being challenged as professional competences and expertise become progressively more complex. This is mainly because problems are more technology-bounded, unstable and ill-defined with the involvement of various integrated issues. Solving these problems requires interdisciplinary knowledge, collaboration skills, and innovative thinking, among other competences. In order to facilitate students with the competences expected in their future professions, educational institutions worldwide are implementing innovations and changes in many respects.

This book series includes a list of research projects that document innovation and change in education. The topics range from organizational change, curriculum design and innovation, and pedagogy development to the role of teaching staff in the change process, students' performance in the areas of not only academic scores, but also learning processes and skills development such as problem solving creativity, communication, and quality issues, among others. An inter- or cross-cultural perspective is studied in this book series that includes three layers. First, research contexts in these books include different countries/regions with various educational traditions, systems.. and societal backgrounds in a global context. Second, the impact of professional and institutional cultures such as language, engineering, medicine and health, and teachers' education are also taken into consideration in these research projects. The third layer incorporates individual beliefs, perceptions, identity development and skills development in the learning processes, and inter-personal interaction and communication within the cultural contexts in the first two layers.

We strongly encourage you as an expert within this field to contribute with your research and help create an international awareness of this scientific subject.

For a list of other books in this series, visit www.riverpublishers.com

MOOC Courses and the Future of Higher Education

A New Pedagogical Framework

José Gómez Galán

Metropolitan University, Puerto Rico, USA
and
Catholic University of Avila, Spain

Antonio H. Martín Padilla

Pablo de Olavide University, Spain

César Bernal Bravo

King Juan Carlos University, Spain
and
University of Almeria, Spain

Eloy López Meneses

Pablo de Olavide University, Spain

River Publishers

Published, sold and distributed by:
River Publishers
Alsbjergvej 10
9260 Gistrup
Denmark

River Publishers
Lange Geer 44
2611 PW Delft
The Netherlands

Tel.: +45369953197
www.riverpublishers.com

ISBN: 978-87-7022-062-0 (Hardback)
978-87-7022-061-3 (Ebook)

Contents

Preface

Information and Communications Technologies (ICT) today imply profound social transformation, one which needs to be engaged by education. One of the main concerns of the education system is the integration of technologies and media resources into the teaching-learning process as well as the need for a critical analysis to be undertaken into the importance of these resources in our world. In this framework, ICT teacher training is fundamental. Alongside these two important issues there is the continuing emergence of new technological proposals which are emerging in the context of the digital paradigm, and are presenting themselves as not only being innovative but also promise to essentially change the meaning of education in the fields in which they are applied. They are having a significant impact and a great amount of experimental practical and scientific studies are being realized especially in relation to the advantages and disadvantages of their potential, which is essential for the definition and assessment of their true significance.

A clear example of this educational-technological impact can be found in MOOC (Massive Open Online Courses) courses. These courses are based on the principles of massive, free access to all materials and resources offered online. This phenomenon has had a major worldwide expansion, opening opportunities at the same time for education and training. In addition to being the entry point for the popularization of science, the future possibilities are enormous and are being studied in all their various dimensions.

The realization of this book is essential because the MOOC paradigm has provoked a great revolution in different parts of the world, and produced enormous change and progress in society. Many initiatives as a result have been developed to implement this new form of education. The courses are open, participatory and distributed along a pathway for connection and collaboration as well as job sharing. Some experts consider these courses to be positive while others see them as a threat to current educational systems. The phenomenon is nevertheless expanding rapidly worldwide, with such speed that the word tsunami is often quoted by some authors when referring to MOOC, It is therefore imperative to understand how it works internally as well as studying strengths and advantages and the undoubted potential it has for the enrichment of teaching and learning in the 21st century (Gómez Galán and Pérez Parras, 2015).

List of Figures

List of Tables

List of Abbreviations

ALISON	Advance Learning Interactive Systems Online
EHEA	European Higher Education Area
EITO	European Information Technology Observatory
GWI	Global Web Index
HDI	Human Development Index
ICT	Information and Communication Technologies
MIT	Massachusetts Institute of Technology
MOOC	Massive Open Online Courses
ONTSI	National Observatory of Telecommunications and the Information Society
OT	Technological Observatories
POOC	Personalized Open Courses
UNDP	United Nations Development Program
WSIS	World Summit on The Information Society

Introduction

Higher education has placed faith in the rise of the MOOC teaching format in a socio-economic context in crisis, and a questioning of the university as a generator of new knowledge and as a guarantor of training and access to the world of work. Also, critical voices emerge concerning the process of harmonization of European higher education, especially due to the lack of internationalization and international mobility among the countries of the European community.

In this macro-context, one might underline that higher education is under-budgeted, with pending structural changes, including the time-consuming process of restructuring the teaching staff and digitization of the institution that has been called, e-university.

Thus, the incursion of MOOCs in the models of the university training proposals, for some, has responded to the universalization of higher education studies, separating the meritocracy from learning, and finally, renewing an institution that demands a new reform to adapt to the neoliberal society.

For others, the MOOCs – attending to the majority type focused on contents – are only a response to the crises faced by the institution, including firstly, the lack of students, secondly, an ideal formative proposal for restructuring the teaching staff, a revision of the degrees, especially those of the second cycle and of the accreditation systems, and thirdly, segregation of the students, with a greater majority abandoning their studies before completion, all understood as a degradation of an individual decision in front of the defense of the very nature of the university itself.

This work has been structured in three sections: the first section covers the university institution in the knowledge society, the second analyzes the MOOC training proposals, and the third discusses the future role of the MOOCs.

This work is born from the collaboration and the experience that the authors have acquired on this subject, with a debate over the ideas expressed in other works of their own and those of other researchers in this field, to whom we are grateful.

1

University and the Knowledge Society

During the last decades, our society has witnessed an authentic revolution that, in an astonishing manner, has deeply influenced, modified, and transformed the way of human existence. This constant and unstoppable revolution is transmuting all areas of our life: social, cultural, personal, labor, economic, training, etc.

This new society is characterized by a high generation of knowledge and the constant, fluid processing of information. In the structure of this new society there is a close interdependence among the different spheres: social, political, and economic. This is precisely the key to understand the new structure of the society (Flecha and Elboj, 2000). Likewise, as noted by Cabero and Aguaded (2003), the inception of information technologies in the daily life of citizens is undoubtedly the most noteworthy occurrence in this millennium.

In the classic book on the information age, Professor Castells (1997) indicates the three features that define this information society: "*first as a basis for a technological revolution; secondly, a profound reorganization of the socioeconomic system, a process known as globalization; third, an organizational change that is no less profound, such as the shift from vertical hierarchical organizations to 'networked organizations*'" (1997, p. 47). The symbiosis and interaction of these three elements generate new socio-cultural phenomena.

We are discussing about a technological revolution that is unprecedented in the history of humanity and that is transforming our communities and cultures into a process that continues with a constant and vertiginous force. These transformations are penetrating and modifying extensively, from the constitutive tissue to the underlying pillars of the society. And in this context, it is the unstoppable progress of the Information and Communication Technologies (ICT) that, directly or indirectly, determines the countless social transformations, experienced by us, which are affecting all areas of development and social progress. (López-Meneses and Miranda-Velasco, 2007).

At present, capital, technology, management, information, and markets are globalized. Globalization, together with information technology and the innovative processes it fosters, is revolutionizing the organization of work, the production of goods and services, relations between nations, or local culture, and even the very foundation of human relationships and of social life (Carnoy, 2009).

In agreement, we share the reflections expressed by Professor Román (2002, p. 17), when he states that it is not a novelty to say that we have gone from an industrial society model to another model called the information society. This society in which we live is nothing more than an evolutionary state of advanced societies in which information replaces the old factors of production and the creation of wealth. Manual work gives way to intellectual work, and power is based, less and less, on tangible assets and more on the ability to self-manage information. According to this author, this society is also marked by the emergence of new labor sectors, the complexity of processes and products achieved immediacy, progress and the constant search for efficiency, the globalization of the media, ideological pluralism, and the multifocality of the community.

In the new society, innovation, change, transformation, and mutability are the characteristics of knowledge generated by people, institutions, universities, companies, or any other human social group. Using the concept proposed by the Polish sociologist Zygmunt Bauman, our postmodern time is a liquid epoch.[1] Faced with solidity, the durability of thought, and the social systems of the past, at present, new ideas, new practices, and new phenomena are constantly appearing that make the knowledge and certainties we possess uncertain and ephemeral (Area, 2009).

In this constant mutability of our reality, the foundations of an authentic social revolution have been forged. As Cabero (1995) comments, the establishment in our society of the so-called "new technologies" of communication and information is generating unsuspected changes if we compare them to

[1]Bauman argues that in liquid modernity, identities are similar to a volcanic crust that hardens, melts, and changes constantly in form. The author argues that they seem stable from an external point of view, but that when viewed by the subject himself or herself, appears in a context of fragility and constant tear. According to his ideas, in liquid modernity, the only value is the need to acquire a flexible and versatile identity that faces the different mutations that the subject has to face throughout his life. More information: http://bit.ly/2nN3KF7.

those that originated the appearance of other technologies, such as printing or electronics (Angulo and Bernal, 2012). Its effects and scope are not only placed in the field of information and communication, but they exceed it to provoke and propose changes in the social, economic, labor, legal, and political structure. And this is because they are not only focused on capturing information but also, significantly, the possibilities of manipulation, storage, and distribution of such information.

According to Castells (1998), the most important characteristics of this new technological society are as follows:

- Technologies are to act on information, to transform it. The information is only the raw material;
- Ability to penetrate the effects of new technologies in most areas of human activity;
- Interconnection of the entire system, through global networks that allow a series of interactive possibilities increasingly complex. This allows to provide structure and flexibility to the technological system, thus allowing organizational fluidity and the convergence and integration of technologies in a general system.

On the other hand, according to Professor Cabero (2003c), this new technological society has the following characteristics:

- Globalization of economic activities, communication, and information;
- Increase in consumption and mass production of consumer goods;
- Replacement of mechanical production systems, by others electronic and automatic;
- Modification of production relations, both socially and from a technical position;
- The continuous selection of areas of preferential development in research, linked to the technological impact;
- Work flexibility and labor instability;
- Emergence of new labor sectors, such as that dedicated to information, and new work modalities such as teleworking;
- To revolve around the media and more specifically around new ICTs, as a hybrid resulting from information technology and telematics, consequentially, the empowerment of the creation of a technological infrastructure;
- Globalization of traditional mass media and interconnection of both traditional and innovative technologies, in order to break the spatio-temporal barriers and reach the farthest;

- The transformation of politics and political parties, establishing new mechanisms for the struggle for power; and
- The establishment of quality principles and the search for immediate profitability in both products and results, reaching proposals at all levels: cultural, economic, political, and social.

We could incorporate other characteristics such as planetarization and the simultaneity of the changes, and their speed. Likewise, as Marquès (2000) points out new challenges for people, among which are the following:

- The continuous change, the rapid expiration of information, and the need for ongoing training to adapt to the requirements of professional life and to restructure personal knowledge;
- Handle a large volume of information and the need to organize a personal system of information sources and have some search and selection techniques and criteria;
- The need to verify the veracity and timeliness of the information;
- The new communicative codes that we must learn to interpret to send messages in the new media;
- The need to verify the veracity and timeliness of the information;
- The new communicative codes that we must learn to interpret to send messages in the new media;
- The tension between the long and the short term at a time when the ephemeral predominates and quick solutions are sought despite the fact that many of the problems require long-term strategies;
- The tension between tradition and modernity: adapt to change without denying ourselves and losing our autonomy;
- Become citizens of the world (and develop a social function) without losing our roots (tension between the global and the local);
- The problems of sustainability at the level of the planet;
- See that the new media contribute to spread culture and well-being in all the peoples of the Earth; and
- Think about the jobs that will be needed and prepare people for them, thus helping avoid unemployment and social exclusion.

We are witnessing, therefore, important social transformations, a good part of which, as already mentioned, are being fostered by the development of different ICTs and by the incorporation of these into society. The spectacular development of ICT has modified the ways of transmitting, classifying, and processing information, as well as the modes of communication and relationship, with a generalized scope on all activities and areas of the human being, from macro and microeconomic spheres, political, social, cultural, labor, or training, even more personal spaces such as family, social relationships, etc. (Gómez Galán, 2003, Guzmán et al., 2004, Castaño and Llorente, 2007, Orellana, 2007, Ponce, Pagán and Gómez Galán, 2018), even on the notion of what is a cultured person (Barroso and Llorente, 2007). As Echevarria points out (2000), new technologies give birth to a new social space, the third environment, which differs clearly from the natural and urban environments.

In this third environment, the processes related to information and communication are of utmost importance. Its influence is such that at the time of naming the society of this century, terms such as "Information Society", "Web Generation", "Generation I" (Internet and/or Information), "Info-society", "Tele-society", "e-Society", "Knowledge Society", "Cibersociety" or, in the words of Professor Manuel Castells (2000), "*Society in network*" or "*Era of information*", and continuing, in this way, reference is made to the changes of a social nature, which are being generated as a result of the use of social software as a means of communication. Perhaps, the term "Information Society" is the one with the greatest strength in the sociological literature, and even in family and social settings. Undoubtedly, information is the basic resource of society in developed countries, characteristically defining the profound transformations of the culture and modes of production that are taking place (Aguaded, 2002).

For Marquès (2001), "the current changing society, which we call the information society, is characterized by continuous scientific advances (bioengineering, new materials, microelectronics) and by the trend towards economic and cultural globalization (large world market, unique neoliberal thinking, technological apogee, digital convergence of all information...). It has a massive diffusion of information technology, telematics and audiovisual media in all social and economic strata, through which it provides us with new communication channels (networks) and immense sources of information; powerful tools for the information process; electronic money, new values and social behavioral guidelines; new symbologies, narrative structures and ways of organizing information ... thus configuring our visions of the world in which we live and thus influencing our behaviors".

On the other hand, Martínez-Sánchez (2007) develops the aspects that, he understands, affect in a significant way this information society (Figure 1.1). Globalization is one of the elements that most attracts attention because of the implicit connotations it contains. A globalization that has generated an economic interdependence, but as we have already mentioned, is not only in this economic plane, but also transcends social, cultural, and

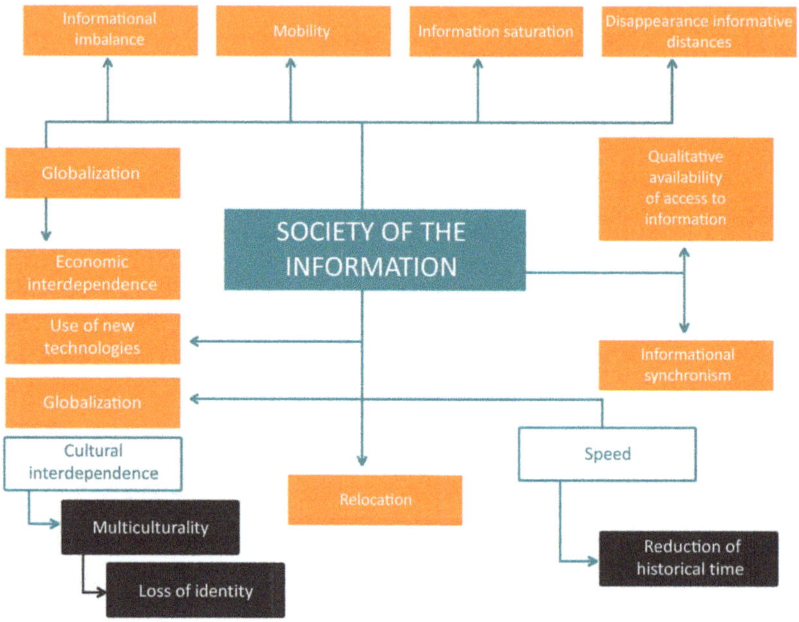

Figure 1.1 Characteristics of the information society.
Source: Martínez Sánchez (2007:5). (Own elaboration)

leisure, and therefore, greatly influences the styles of people's lives. This author also influences issues such as information overload or information saturation, fostered by factors such as globalization and off-shoring, the qualitative increase in the capacity of access to information and information synchronism that allows us to be constantly informed of any event when event occurs, and all this thanks to the near ubiquity of the media through the use of new technologies.

Following Cabero (2008, p. 20), so as to say, the information society proposes to do it around a series of circumstances that, to a greater or lesser extent, affect all the individuals equally:

- The globalization of economic activities, and therefore of the Society, not only in relation to the economy, but also in relation to culture, leisure and lifestyles;
- The concept of time and space are being transformed;
- Presence of ICT in all key sectors of society, from culture, to business, without forgetting education;

- Another characteristic is the "learning to learn", whose incorporation is not stressed equally in all places, so that there is an important digital divide that is causing social exclusion;
- It has gone from a society based on memory to the knowledge society;
- That it is a society characterized by complexity and dynamism;
- And that it is a network society, not of individuals or isolated institutions, but of individuals and institutions connected in virtual communities.

Parallel to the concept of the information society, we find the "Knowledge Society", a concept that stands in relation to the management and creation of knowledge in companies and organizations, a more academic term, as a more integral alternative to the concept of the information society, which is not only related to the economic dimension but also the idea of technological innovation. This concept refers to the unprecedented acceleration in the rate of creation, accumulation, distribution, and depreciation of knowledge (Vargas, 2005); to that society based on the critical, rational, and reflective use of global and distributed information (Gisbert-Cervera, 2000); to the alteration in the accumulation and transmission of knowledge, many of which are new or concealed in contexts distant from those where they were born (Steinmueller, 2002). It is a society wherein the conditions of knowledge generation and information processing have been substantially altered by a technological revolution centered on information processing, knowledge generation, and information technologies (Castells, 2002). It refers to human resources displacing natural resources as key inputs and sources of competitive advantages, knowledge as a key variable of power in society (Tedesco, 2000), and the prominent use of information technologies in the generation of knowledge. This society is characterized by an economic and social structure, in which knowledge has replaced work and asserts raw materials and capital as the most important source of productivity, growth and social inequalities (Drucker, 1994 y 2017).

As Kruger (2000) points out, the knowledge society is characterized by three trends: the intensive use of ICTs in all social areas, the globalization of economic processes, and the emergence of a scientific-technological civilization. This connects with the works of Pérez (2004) wherein he identifies five basic features of the knowledge society:

(a) the intense production of knowledge;
(b) its transmission through education and training;
(c) use of human capital in productive activities;

(d) the accelerated dissemination of information through ICTs and their networks; and
(e) the economic exploitation of knowledge through innovation, especially in the productive sectors with the highest technological content.

In short, as can be seen from the aforementioned, our society is undergoing profound transformations and radical changes in its structure. To make this situation a reality, they have contributed unequivocally and manifest the ICTs, and more particularly, the Internet (López-Meneses and Miranda-Velasco, 2007).

In a relatively short time, the Internet has become the largest universal information library, a kind of an extensive virtual media library in permanent expansion and updating itself constantly. The internet offers the opportunity to access services that can help develop our work, facilitates our daily tasks, and allows us to enjoy our leisure time (López-Meneses and Ballesteros, 2000; Gómez Galán, 2017a).

Ortega (1997) points out that the Internet is a socio-cultural phenomenon of growing importance, a new way of understanding communications that is transforming the world, thanks to the millions of individuals who daily go to this inexhaustible source of information.

The development of the Internet seems to bring, along with other changes of a different nature, the possibility of a profound transformation in the field of interpersonal communication and, in general, in all the processes of information flow (Cañal, Ballesteros and López-Meneses, 2000). Besides, it also facilitates and strengthens the construction of collective knowledge communities (Valverde and López-Meneses, 2002).

In the same discursive line, the Network has progressively changed from being a repository of information to becoming a social instrument for the elaboration of knowledge (Cabero, 2006a). Every new development is published on the web, deepening the collective knowledge and widening the technological capacity of the network, thereby making it easier to use (Castells, 2001). In this sense, the possibilities of exchanging information among the scientific community are greatly expanded, achieving immediacy in the publication of the results, lowering production and distribution costs, and favoring the presentation of ideas and research results, regardless of the location of the subject.

Sanz (1995) formulates that, from a broader point of view, the Internet Network constitutes a socio-cultural phenomenon of increasing importance, a new way of understanding the communications that are transforming the world, thanks to the millions of individuals who have an everyday access to this inexhaustible source of information (the largest that has ever existed), resulting in an extensive and continuous transfer of knowledge among them.

The network makes it possible to universalize information, communication, and knowledge with immediate accessibility. Images that until recently could be considered science fiction are, every day, more usual in our immediate context. Today, nobody is surprised that some people go talking to others through a mobile phone or watching television using it and also that direct communication can be made in real time with images and sounds with people located in very different points of the continent or the planet (Cabero, 2006b).

In this sense, as pointed out by Barry Wellman (2004)[2] in "*The Internet in Everyday Life: An Introduction*", the Internet is affecting the traditional forms of sociability; gradually, users spend more time browsing, use e-mails extensively, and acquire goods through e-commerce, and there are more discussion groups. In short, the Internet is reaching higher levels of democratization and is gradually entering domestic life as a fundamental instrument that positively affects studies, work, and communication, being a part of everyday culture in advanced countries.

The Internet has become an essential element of our culture, but to what extent this is so, if the data published by the Internet World Stats website is taken as a reference[3] in relation to the worldwide use of the Internet. In this sense, if we analyze the percentage of Internet users by regions (Figure 1.2), we can verify that half of them (50.2%) are found in the Asian continent, followed by Europe (17.1%), Latin America and the Caribbean (10.3%), and Africa with 9.3%. North America (8.6%), the Middle East (3.8), and Oceania (0.7%) occupy the last three positions.

Apparently, it might seem that, indeed, the rate of Internet penetration worldwide is considerable, and there are no major geographical differences. It highlights the fact that in regions such as Asia, Africa, or Latin America and the Caribbean, where the greatest number of countries with a Human Development Index are concentrated.[4] Lower, are higher in percentage of

[2]Source: http://www.chass.utoronto.ca/~wellman/publications/index.html

[3]Internet World Stats. URL: http://www.internetworldstats.com

[4]Human Development Index (HDI). INDICATOR created by the United Nations Development Program (UNDP) in order to determine the level of development and have countries base their evaluations, not only on the income of people in a country, but also to consider if the country brings its citizenship to grow in an environment where they can develop their learning under optimal living conditions.

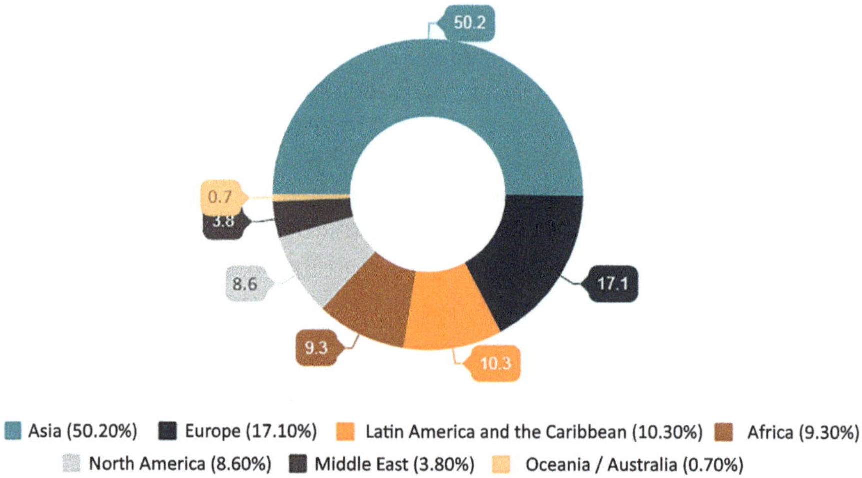

Figure 1.2 Internet users in the world by regions.
Source: Internet World Stats. (Own elaboration)

Internet users than regions and countries with a higher rate of development, such as North America, or Europe. However, if we carry out a broader analysis and compare these data to the percentage of the Internet user population, provided by the International Communications Union,[5] with the estimated population data for 2017 provided by the United Nations[6] (Table 1.1), we can see that there is certainly a spectacular growth in the number of Internet users in the last 17 years (from 2000 to 2017) in the areas previously mentioned (Africa with 7,557.2%, Middle East with 4,220, 9%, Latin America and the Caribbean with 2,035.8%, and Asia with 1,539.4%). However, as shown in Table 1.1 and Figure 1.3, these zones present, at best, penetration rates slightly above the world average (49.6%), as is the case with Latin America and the Caribbean. But the Caribbean (59.6%) or the Middle East (56.7%), in other cases, have much lower rates, as in Africa (27.7%) or Asia (45.2%). Therefore, these regions are still far from achieving penetration rates comparable to the regions with greater development such as North America (88.1%), Europe (77.4%), or Oceania (68.1%).

[5]ITU – International Telecommunications Union. URL: http://www.itu.int/es/Pages/default.aspx

[6]United Nations. Department of economy and social affairs. Population division. URL: http://www.un.org/en/development/desa/population/

Table 1.1 Statistics of population and Internet users in the world (March 2017)

Regions	Population 2017 (Estimated)	% World Population	Internet Users Mar 31, 2017	Penetration Rate	Growth 2000–2017	% Users
Africa	1,246,504,865	16.6%	345,676,501	27.7%	7,557.2%	9.3%
Asia	4,148,177,672	55.2%	1,873,856,654	45.2%	1,539.4%	50.2%
Europe	822,710,362	10.9%	636,971,824	77.4%	506.1%	17.1%
Latin America and the Caribbean	647,604,645	8.6%	385,919,382	59.6%	2,035.8%	10.3%
Middle East	250,327,574	3.3%	141,931,765	56.7%	4.220.9%	3.8%
North America	363,224,006	4.8%	320,068,243	88.1%	196.1%	8.6%
Oceania/Australia	40,479,846	0.5%	27,549,054	68.1%	261.5%	0.7%
Total world	7,519,028,970	100.0%	3,731,973,423	49.6%	933.8%	100.0%

Source: The Internet World Stats. (Own elaboration)

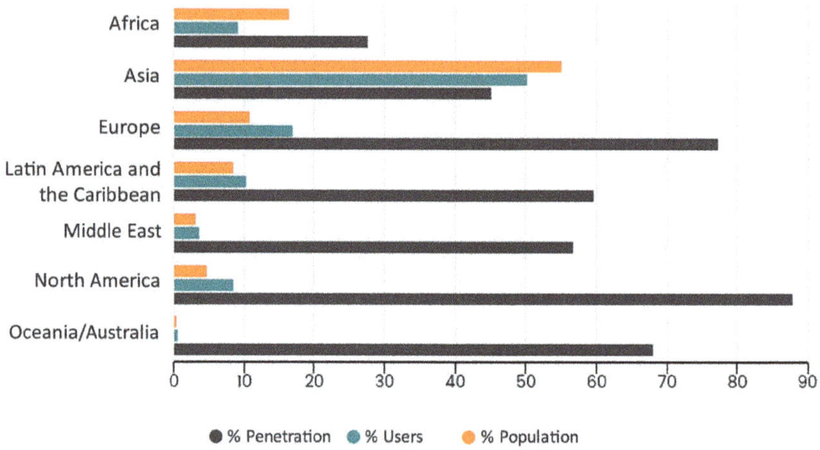

Figure 1.3 Comparison between percentage of population, users, and Internet penetration rates.
Source: Internet World Stats. (Own elaboration)

We understand that these data, although not very favorable in terms of total access to the Network, should not be taken at all with pessimism, since the rate of growth in percentage of the population with Internet access in these most disadvantaged areas. It is certainly spectacular and, therefore, quite hopeful. However, we must not ignore the reality of these data. While in the developed countries, a high penetration rate of Internet access has been reached, in the developing countries only a small percentage, close to 50% of the population (and in some cases even lower), have access to the Internet.

This situation of marginalization on the Net has direct consequences on the separation between the people and the countries and has an impact on increasing social exclusion. As Estefanía (2003, 10) indicates: “Inequality feeds on the richness of the system. As progress is made in the technical and economic levels, not in all, the social aspect falls back.” In the words of Yoshio Utsumi, former Secretary General of the ITU (2003), “ICTs, including the Internet, are creating many new opportunities, but as a consequence of their uneven expansion, they are also creating new challenges, particularly the emergence of new technologies, ‘digital gaps’. Leaders from around the world should guide the evolution of the information society and create a more just, prosperous and peaceful world.”

The Digital Divide concept, according to Cabero (2004), refers to the differentiation produced between those people, institutions, societies, or countries that can access the Network and those that cannot. In other words, there is an inequality among the possibilities of accessing information, knowledge, and education through the new technologies. While talking about the digital divide, we are referring to the great inequality that arises in societies due to the difference between the population that could access ICTs and incorporate their use in their daily and routine life, and among that part of the population that does or does not, one has access to the technological advances or, if one has effective and functional access, one does not know how to use them (Ballesteros, 2003, Serrano and Martínez, 2003).

Access to information and knowledge has become an important tool for countries and social groups to evolve to better levels of development. However, the risk of this situation is that this digital divide is becoming an element of separation, of e-exclusion, of individuals, collectives, institutions, and countries, and thus the separation and marginalization, merely technological, is becoming social and personal separation and marginalization. In other words, the digital divide becomes a social gap, so that technology is an element of exclusion and not of social inclusion. Hence, we cannot speak of a single digital divide, but of different situations of e-exclusion.

In this sense, Castaño (2008) presents a classification of digital divide, according to which it is possible to speak of the **first digital divide**, related to access to the computer and Internet connection in relation to the socio-demographic characteristics of individuals (age, sex, studies, etc.) and, on the other hand, the **second digital divide**, the one that affects the uses (both intensity and variety of uses) and is determined by the abilities and skills of individuals to use technological devices and access the Internet. This second gap tends to affect mostly women, because of what is used in most studies on gender digital divide. According to this author, the most complex challenge refers to use and skills, relating to the concept of quality over that of quantity.

Therefore, this situation is not just a matter of presence or absence of ICT; it is a question of degree, of the separation between one country and another, or between different groups. The digital divide needs to be measured not only in terms of the number of telephones, number of computers, and Internet sites, but also in terms of options, facilities, and adequate costs for access to the network and training and education programs to optimize the use of the infrastructure created. Reducing the digital divide by implementing telecommunications and information infrastructure will not necessarily reduce the socioeconomic disparity. The reduction of the digital divide will be possible as long as material, intellectual, and moral education initiatives are carried out to ensure its continuity and sustainability (Cabero, 2004, Serrano and Martínez, 2003).

Regardless of these troublesome situations of technological exclusion, we are certainly at a historic moment, in which, for the first time, there are access roads to information and knowledge that, with a decreasing cost, allow greater access to different people and social groups. Although this fact, in itself, as can be elucidated from the above, does not necessarily imply a greater democratization, it does establish a dynamic that is oriented towards decentralization, something new until now in technological development, always tendentially centralizing everything. It is a qualitative change that makes a space possible for participation, non-hierarchical social control, and self-management.

In this non-physical space wherein the Internet is located and human relationships are increasingly developing, also defined as the cyberspace or "*Third Environment*", along with the unquestionable and enormous advantages it presents, we can discover inequalities, such as those previously referred to, in relation to access. But in addition, we find authentic attacks on human rights related to the limitation of access to technical, economic, or cultural conditions that would allow the development of more advanced forms of public participation and exchange and free expression of ideas and beliefs (Echeverría, 1999).

In advanced countries, the powers that we are trying to underline the dangers and abuses in the use of information and communication technologies, especially in relation to morality, in order to look for reasons that support the establishment of limits. Totalitarian regimes directly censor access or dissemination of any content considered contrary to their interests. In these, the limitations on human rights are exercised by governments that tend to be concerned with the ability of new media to promote the creation of horizontal organizations of their governed, freedom of expression and the construction of virtual social communities that escape one's control. And given the characteristics of cyberspace, these limiting actions that infringe rights are, at times, hardly observable, and evident only to some in the community.

For this reason, according to Bustamante (2001), the Internet appears as one of the scenarios where one of the most decisive battles for freedom of expression is resolved, and therefore, for human rights in general. Consequently, it is especially relevant to talk about the condition of rights in the new digital 2.0 environment, and the attacks that citizens of the global village can suffer through communication and information technologies.

In this sense, the author takes a journey through the history of human rights that culminates in the new formulations of these rights as a result of the emergence of the technological society. This author establishes four *generations* in the evolution of human rights:

1. *First generation human rights.* They have to do with the right to *freedom* of individuals: the right to dignity of the person and their autonomy and freedom against the state, their physical integrity, or procedural guarantees. These are rights that are supported by the philosophy of enlightenment and the theories of social contract. They also come from the liberal constitutionalist tradition. These rights are included in the *Universal Declaration of Human Rights* (1948) and the *International Covenants* on Human Rights (1966–67), which include, among others, Civil, Political, Economic, Social, and Cultural Rights.
2. *Second generation human rights.* They are economic and social and affect the expression of *equality* of individuals. These rights are incorporated from a tradition of humanist and socialist thinking, which requires some intervention by the state to guarantee equal access to first-generation rights. The state is requested to guarantee access to education, work, health, social protection, etc., thereby creating the social conditions that enable a real exercise of freedoms in a society where not all human beings are born equal.
3. *Third generation human rights.* They emerged in the second half of the 20th century under the name of solidarity rights: the right to privacy, the right to enjoy clean air, the right to receive solid information, consumer rights, the right to protection of heritage, the right to development, and the right, in general, to have access to a quality life. These rights have been gaining an increasingly important role, and thanks to them, the concept of North–South dialogue has been developed, the respect and preservation of cultural diversity, the protection of the environment, or the conservation of the cultural heritage of humanity.
4. *Fourth generation human rights.* The phenomenon of globalization, the transition from the *Information Society* to the *Knowledge Society*, the integration of the planet through the universal extension of mass media, as well as the phenomena of multiculturalism caused by migratory flows, are symptoms that something substantial is changing. Today the right to peace and intervention in armed conflicts is based on a legitimate international power; the right to create an International Criminal Court that acts in cases of genocide and crimes against humanity; the right to sustainable development that allows preserving the natural environment and the cultural heritage of humanity; the right to a multicultural world wherein ethnic, linguistic, and religious minorities are respected, or the

> right to free movement of persons, not only of capital and goods, which allows decent living conditions for immigrant workers. On the other hand, the possibilities that open up from the omnipresence of technology in social life are so many that a new ethic demands a more global and imaginative protection of the rights of individuals. These rights would be encompassed in what could be considered a *fourth generation* of human rights, in which the *universalization of access to technology, freedom of expression on the Internet, and* the *free distribution of information,* play a fundamental role.

In relation to this fourth generation of Human Rights that Bustamante (2001) tells us about, at the World Summit of the Information Society held in 2003, organized by the International Communications Union and sponsored by the United Nations, it was concluded that the Information Society should be a social, cultural, economic, and political space of equal opportunities for access to information resources, in which, together with digital information and communication technologies, a state of generalized digital inclusion was generated, in other words, a space where all citizens had access to information networks on equal terms and knew how to use their instruments[7] (ITU, 2003). Likewise, this "inclusion" expresses a human right that would involve access to ICT at least in the following terms:

1. Availability and development of technology appropriate to the reality and need of the communities to which it is addressed: Universal access to ICT.
2. Social management of technological resources that includes the following:
    - The citizen's right to evaluate the management of the state or its concessionaires in the field of ICT;
    - Regulations to encourage free competition in the field of ICT-related services, guaranteeing equal opportunities for all providers of small and large services;
    - Development of participatory and community management models in ICT matters, including consultation processes for the adoption of public policies;
    - The training in the communities to develop themselves as promoters and multipliers of ICT and raising awareness through strategies that allow people to know and apprehend the phenomenon and use of ICT.

[7] *Source*: World Summit of the Information Society (2003). URL: http://bit.ly/2pckTMq

3. Comprehensive training regarding ICT appropriate to all levels and social groups.
4. Promotion, development, and freedom of circulation of one's own contents and knowledge processes about ICT.

With regard to the European level, various studies have found that some situations of exclusion are not the sole or direct causes of situations of economic or material deprivation, but deprivation of human rights and citizenship. Due to this, the strategies of the European Union for the fight against exclusion do not focus as much on reducing poverty as on the implementation of plans that guarantee social inclusion, making this a transversal objective of any political initiative (Cobo, 2017).

It is clear that we still have a long way to go in the socio-educational fields towards the development of a new ethic to analyze and deepen the solidarity uses of power that ICTs put in the hands of states and their citizens. But hopefully, this network of networks can become in a not too distant future a tool that guides the general public, that it is a means to help change, innovation, transformation, and social integration, as well as a vehicle for democracy and equality among the population and, ultimately, to become a true universal service of public participation, exchange and free expression of ideas, beliefs, and a citizen space of solidarity for the welfare of the village globalization and the intercultural training of Homo Digitalis.

However, before this approach, a great challenge for society arises, since this situation demands the need for teachers with innovative and creative ideas, capable of adapting to the multiplicity of situations that contemporary life is designing and that can participate in the social, cultural, and educational transformation that the rapidity of technological development demands (Gómez Galán, 2003 and 2015).

At present, the advancement of technology and the processes of change and innovation described lead progressively in developed countries to a new type of society wherein working conditions, leisure, relationships, and communication, or mechanisms of information transmission, adopt new forms. All these transformations cause a change in individuals. Hence, their preparation and education are also being affected, requiring a reformulation in terms of content and forms, which leads to the thinking that a new educational approach is necessary (Gallego, 2004).

It is no longer enough to know, but also, it is necessary a knowledge linked to the profound economic and social changes under way, with new technologies, with the new industrial and institutional organization, in an increasingly complex and interdependent world, which requires people with alive creative and innovative restlessness, with a critical, reflective, and participative spirit (Ballesteros, López-Meneses and Torres, 2004).

In this sense, the declaration of the World Summit on the Information Society (WSIS, 2003) reminds us that, among other principles, it is recognized that education, knowledge, information, and communication are essential for progress, initiative, and the welfare of human beings. Moreover, ICTs have immense repercussions in practically all aspects of our lives. The rapid progress of these technologies offers unprecedented opportunities to achieve higher levels of development. The ability of ICTs to reduce many traditional obstacles, especially time and distance, makes it possible, for the first time in history, to use the potential of these technologies for the benefit of millions of people around the world.

As a result of the foregoing, the "Technological Revolution" has given rise to the "Information and Knowledge Society". Under this scenario and using the words of Herrera (2003), we can consider that living in the current era can be approximated to change, or at least to an adaptation to the continuous changes that society will experience. This implies that we will need to be in constant formation throughout our lives.

On the other hand, although, networks have a design and structure that respond to a will consciously oriented to the promotion of a democratic means of free expression that can allow the development of more advanced forms of public participation, exchange and free expression of ideas and beliefs, too, can be the creator of serious dangers (cultural uniformity, social exclusion, and increase in educational inequalities), especially in those people who are not able to adapt to the demands of living in it, either through disinterest, ignorance, or also due perhaps to the lack of economic resources, or a lower educational level.

Thus, as Cabero et al. (2002) state, this new *Information Society* is a challenge for teaching, requiring educational actions related to the use, selection, and organization of information so that the student is trained as a mature citizen. Perspectives to consider are allowing flexible teaching structures necessary in the new conception of the teaching and learning process, where the student has an active participation, attending to their emotional and intellectual skills and assuming responsibilities in a fast and constant world that allows the student to enter into a working world that will demand lifelong training. Being one of the characteristics that calls attention to this new technological society is the idea of efficiency and constant progress.

Not in vain, it is more than evident that we are submerged in a perennial change towards a new technological and cultural era, which implies a new fact of civilization governed by the nature of a technological innovation, unpredictable and discontinuous (Ballesteros and López-Meneses, 1998; Gómez Galán, López Meneses and Cobos, 2016).

In this way, the social changes that have taken place in Europe in particular, and in the whole of Western societies in general, have been very profound and have affected intensely the labor market, the economy, and lifestyle of the citizens. These techno-social and economic circumstances are located in the emerging information and knowledge society accompanied by two educational profiles, key to its proper community development: the principles of "dynamic learning" and "lifelong learning". Both are linked transversally to the incorporation and use of new information and communication technologies and critical and responsible participation to adapt to emerging social and technological trends. And, it is obvious that for the complete development of the previous principles, it is of vital importance the daily teaching task of the professionals of the education process as facilitators and managers in social intervention and as social agents for the technological alphabetization in a diversity of contexts and situations.

2

The Use of Digital Technology in University Teaching and Learning

The university has been undergoing changes, in terms of the concrete expression of the concept of science and knowledge as a continuous dialogue with postmodern society, with the consequent modification that it implies in its organizational structure and in its methodological habits. For quite some time now, people have been arguing about these needs, from different perspectives (Marcelo, 1993, Ferreres, 1997, Fernández Sierra, 1996; Beraza, 2002). However, it was only in this last decade, with the emergence of ICT, that this need has been confirmed (Gómez Galán, 2014a and 2016a).

The university, at the beginning of the second decade of the 21st century, cannot work under the idea that scientific knowledge is its property and only it has some authority to disseminate it, when daily life shows that access to knowledge/information is found within the reach of anyone in the Network. In addition, the concepts of continuous formation throughout life and self-training are already installed in Western cultures, and it makes for another kind of university student; together with the dynamics of globalization, it forces a different work perspective, in terms of content and methods, as will be outlined below and as well how it is included in the proposals of international organizations.

During the last century, the development of capitalism left two great consequences in the university life: on the one hand, the tension continued in relation to the type of demand that society places in the university and, on the other, the arrival of the middle classes at the level of university studies, with the consequent overcrowding of the system and the classrooms. Regarding the first of these issues, there is a contradiction between the effort for teaching and research that serves the immediate future of both the subjects involved and the societies and, on the contrary, the pressure that the industry, the governments and, even students themselves, to achieve short-term results, measured with immediacy. It is a tension that seems to be increasing as the characteristics of the community are transformed into what is being called the "knowledge society" and which is reflected, for example, in the continuous modifications of the curricula today.

Adapting to new ways of life and new information technologies seems essential, but care must be taken to maintain a balance between the social demands based on the development of the subject and those supported by the expansion of the market. If economic competitiveness was to take precedence in decision making, people would no longer go to the university to learn, but to "buy" a profession (García Roca y Mondaza, 2002, 13). Perhaps it is an exaggeration to put it this way, but given the weight that the market is acquiring in personal, national, and transnational relationships, it is important to ask whether the university's offer should only adapt to the demand, or if it should induce new demands. Because, as these authors affirm:

"Certainly the university needs to recover its sense and its mission in contact with the express demands of society. But it must also feed itself, since knowledge is an 'input' of knowledge; teach to have a critical thinking, instead of being a machine to compete in the market" (García Roca and Mondaza, 2002, 13).

Regarding the second of the issues mentioned above, the change in the type of students, on the one hand, the principle of education throughout life has made the students diversify. Jonasson (1999) classifies them into four major groups:

- Traditional students of the first and second cycle, in initial training (in our opinion, we would have to add those who, also in initial training, extend until the third cycle to obtain the maximum number of possible accreditations for their first insertion in the world of work).
- At first, older students, often simultaneously studying part-time, with a work activity, but from the 2007 Law of harmonization of the European higher education system, it is difficult to be able to combine work and university studies.
- Students who have already obtained a first diploma of higher education and complete with third cycle studies, demanding specialized research for advanced professional development.
- Students who, immersed in processes of specialization and professional diversification in areas such as computer science or management, for example, approach new professional horizons by researching general courses related to these domains.

On the other hand, the massive increase in the number of students in the university at the end of the twentieth century caused, in the first instance, that it fulfills functions other than access to knowledge or research and preparation for employment. Some that stood out and that currently could not be denied would be, above all, the function of social mobility and the "park" (Young, 1993) for young people to prepare them and delay their incorporation into the world of work, or "unemployment" (Amando de Miguel, 1979).

This initial phenomenon of massification had fundamental effects on the reality of the university, which still remain and some have increased; thus Zabalza (2002, 28) raised these questions, and here we complete with:

(a) The arrival of increasingly heterogeneous groups of students, not only in terms of ages and interests, as indicated above, but also with regard to intellectual capacity, academic preparation, motivation, expectations, economic resources, etc. There were also changes in terms of open access to women, the incorporation of people seeking complementary training to their work, and in short, a great change in the type of students and their availability to study.
(b) The need to hire new faculty to initially attend the avalanche of students and later, in the first decade of the 21st century, the drastic reduction in spending, especially in the departure of personnel, initially implied consequences for the type of hiring, in the requirement of minimum training requirements and in the possibility of organizing training systems for the best teaching and research; for example, there are knowledge areas with more than 80% of their teaching staff with temporary and hourly contracts. Then one must consider a second phase, with the creation of new contractual figures even more precarious and still with a shorter period of training as interim substitute professors, or other figures that are born of budget reduction and the freezing of public procurement.
(c) The appearance of subtle differences in the status of the studies and the university centers that teach them. Overcrowding has not occurred equally in all the degrees; some of them, such as Medicine, Engineering, etc., have retained their elitist character and, with it, their status as privileged studies. In the degrees of humanities and social studies, massification has influenced the most, multiplying the specialties, maintaining teaching to large groups, incorporating teachers in a way that is also almost massive, and additionally of course, in precarious working conditions.

(d) At the same time, processes of certification, accreditation, and evaluation of higher-education institutions and their services, of degrees and their teaching programs, of teachers and students by virtue of competitiveness, have been imposed.
(e) Finally, and this will be seen later, the digitalization processes of the university and of higher education.

More than a decade later, the European framework of higher education, or what was called “Bologna Plan”, was established and implemented; however, much earlier in 1998, with the Declaration of the Sorbonne in Europe, a process to promote harmonization among national systems of higher education was initiated. This Magna Carta of European universities raises four basic concepts for them:

1. The production and transmission of culture in a critical manner;
2. The dissociability of teaching and research;
3. The freedom of research, teaching, and training; and
4. Ignorance of any geographic or political border.

The Ministries of each member country of the union endorsed, with the signing of the Bologna Declaration (1999) and the subsequent ratification of Prague (2001), the importance of a harmonious development of a European Higher Education Area. The Bologna Declaration included among its objectives the following:

– The adoption of an easily readable and comparable system of qualifications, through the implementation, among other things, of a Diploma Supplement (a kind of annex that translates the characteristics of the curriculum studied by the EU into another EU language) regarding the students, their qualifications, professional profile, etc.;
– The adoption of curricula based on two main cycles, undergraduate and graduate (what is known as “the discussion of 3 + 2 versus 4 + 1”, alluding to the number of years of both cycles). The degree awarded at the end of the first cycle will have to have a specific value in the European labor market. The second cycle will lead to obtaining a Master and/or Doctorate, as already occurring in many European states;
– The establishment of a credit system, such as the ECTS system;
– The promotion of European cooperation to ensure a level of quality for the development of comparable criteria and methodologies;
– The promotion of a necessary European dimension in higher education with particular emphasis on curriculum development. This objective would affect the de facto contents of the subjects we teach, since many

of them should have thematic descriptors that reflect the "European perspective" of the discipline in question.

Supporting this process of convergence achieved in Bologna/Prague, the Parliament and the European Council designed other exchange and mobility programs among the member countries, in addition to programs for the development of second and third cycle degrees, in order to improve the quality of higher education and the promotion of intercultural understanding through cooperation with third countries. It is intended, therefore, to offer "*an authentic and attractive teaching, with international projection, particularly at the level of the third cycle*". Its objectives are:

(a) Promote the emergence of a properly European higher education offer that is attractive both in the European Union and beyond its borders;
(b) Encourage greater international interest in the acquisition of European qualifications and/or experiences among highly qualified graduates and academics from around the world and enable them to acquire these qualifications and/or experiences;
(c) Consolidate a more structured cooperation between the European Union and third-country centers and greater mobility to third countries within the framework of the European study programs.
(d) Improve the profile, visibility, and accessibility of the European education.

In order to achieve these goals, one must move toward conducting masters in the European Union, a system of grants, partnerships with centers and superior engaged third countries, and also ensure proposed technical support measures. These purposes, lines of action, and specific programs, prepared and agreed on, or approved by our political representatives in Europe, have marked the evolution of our universities in the last decade, but we cannot say openly that we are better.

A complex report (CRUE, 2016), centered on Spain but presented common problems in Europe, shows that access to higher education is lower than the OECD average, and although in the grades, the difference is smaller than 54% to 47%, compared to the Master, is a worrying difference of 23% to 9%; this difference can be partly explained by the fact that Spain continues to maintain the highest university public prices in the EU, among the five highest in the European Union. A situation that will be aggravated, on the one hand, when the 3 + 2 model is generalized, that is to say, three years of degree and two of master's, because as we know, Spain decided for the option 4 + 1 at the time, and that now must go through changes slowly, because it will only apply to new degrees, according to the president of the CRUE. On the other hand, if the amount of grants and scholarships is not increased, Spain has reduced by 13.6% in terms of GDP, which is a third of the support of the rest of the OECD countries.

Another aspect that helps explain the uneven evolution of the degrees and masters in the period (2008–2015), according to the same report, is the economic crisis during that period, and therefore the general lack of funding of the university system, one of the aspects which will relapse later, and in the search for more education and training in preparation for access to the workplace. Thus, according to the data, the grades suffered a decline of 4.2% in that period, while the masters increased by 239.2% the number of students in the 2015–2016 academic year, in terms of percentage only 12.9% of the undergraduate students, which, as we have seen before, is still lower than the OECD average.

In the same report, it is shown that the mantra about which the Spanish populations, or the younger ones, are over-qualified with respect to population, is simply not true. Thus, among the youngest (25 to 34 years old) with higher education, there is a difference with an OECD average of 8 points, and with the population in general, in relation to the best university systems in the world, such as the United Kingdom and the United States, of 9 points.

Finally, one might emphasize in favor of the university professors, which despite the generalized reduction in the period (2008–2015) of the autonomous funding of the public universities, plus the freezing of the renewal of teaching bodies, subject to the rate of replacement, etc., is a situation that is positive. In terms of scientific production, this has been increasing from 2.8% in 2000 to 3.3% in 2015; and all this thanks to the efforts of the university professors (both personal and families), who have to support the expense of the research and the dissemination of knowledge.

3

The Virtualization of Teaching in Higher Education: e-University

The university has not been alien to the information and communication technologies (ICTs) taking place in the society; one could speak, for example, of e-university, as well as e-administration, and e-health. A drift has occurred as a result of the trend in all institutions and social services, which have been implementing IT solutions to the mandate of the globalized economy, mainly in the search for efficiency in the provision of services, user management, and cost reduction.

All this is possible due to the evolution of the Network of networks – web2.0 and web3.0 – the development of mobile devices, and the infrastructures that allow the connectivity of people and things and are changing our experience under multiple aspects: in leisure, personal communications, learning, at work, etc. Using the metaphor of Bauman (2006) to characterize the current processes of a socio-cultural change, driven by the omnipresence of the ICTs, this suggests that the current time – the digital culture – is a flow of information and knowledge production unstable, in permanent change, in constant transformation, as opposed to the developed cultural production, mainly in the West throughout the nineteenth and twentieth centuries, where the stability and inalterability of the physical, the material, and the solid, prevailed. In other words, a digital approach is a liquid experience well differentiated from the experience of consumption and the acquisition of solid culture (Area and Pessoa, 2012).

Along the same line of argument, different authors (Gómez Galán, 1999, Marquès, 2000, Cabero 2001, 2003a and 2003b, Martínez-Sánchez, 2007, Orellana, 2007, Sevillano, 2008, Vázquez and Sevillano, 2011, Martín and López-Meneses, 2012; Floridi, 2014; Hajnal, 2018) indicate that the spectacular development of ICTs has modified the ways of transmitting, classifying, and processing information and modes of communication and relationship, with a generalized scope on all the activities and areas of the human being. Martínez Sánchez (2007) develops the aspects that for their consideration affect significantly the information society. The panorama that this author points out is controversial and loaded with meanings, since of all the aspects indicated, globalization is the most striking, because of the connotations that it entails. We speak of an economic, social, cultural, and even of leisure globalization, and therefore of the lifestyles of the subjects, of a globalization of mass media, etc. Globalization affects, then, the technological development of the advanced countries, we speak of computer, audiovisual, and telecommunication technologies; hence, the current social model of past times sees its pillars begin to oscillate, as this information society modifies the way of doing, producing, and receiving the information and also as the subjects participate in its elaboration as well as in its selection and classification processes.

García-Ruiz (2012) indicates that globalization, the expansion of new technologies, and the economic, political, and social changes that have been derived from the society have enhanced the educational debate and the questioning, on the part of postmodern academicism, about the suitability of traditional approaches and paradigms in education.

Regarding the educational field, as pointed out by Hinojo (2006), the current society needs flexible organizational structures in education, which allow for both a broad social access to knowledge and a critical personal training that favors the interpretation of information and the generation of one's own knowledge.

Traditionally, university education has been based on a methodological model focused on the teacher, with emphasis on the transmission of content and its reproduction by students, the master's lesson and individual work, but teaching through ICT currently demands – both on the part of the teaching staff, as well as on the part of the students – a series of changes that generate a rupture of this model, but that at the same time help us guarantee an advance towards the very quality of it (Aguaded, López-Meneses and Alonso, 2010a and 2010b).

Also, as pointed out by Oztok et al. (2013), in the Information and Communication Society, asynchronous and synchronous communication in university contexts should be encouraged, from the role of university teaching staff, through the use of virtual training platforms, so that the learning process is nourished by both personal and academic aspects that configure a learning community.

3.1 Training and Professional Development at the e-University

University professors in the social context presented above require developing new professional skills in the field of practical exercise and their area of expertise; and it is the case that also research in educational institutions can adapt to the current and future contexts.

The initial training of educators is extremely important to be able to face the challenges of the ICT and the social demands in the training institutions (Gómez Galán, 2009, Ruiz and Sánchez, 2013).

According to Cabero, López-Meneses, and Llorente (2012), teachers begin to require an important change of mentality in relation to the traditional modes of action, seeking greater diversity in the methodological procedures and evaluators to incorporate them into teaching practices. Although it is true, this transformation requires a strong institutional commitment that supports and safeguards innovations (Aguaded, Muñiz and Santos, 2011; Gómez Galán, 2016b).

On the other hand, ICT applications, arising from web 2.0, intrinsically involve an active participation of users, who are not only recipients of information and knowledge but also become producers and reconstruction of these. All this is strongly linked to the strategic lines of convergence in the European Higher Education Area (EHEA), which is committed from a pedagogical point of view to collaboration, teamwork, the free dissemination of information, and the generation of its own contents. (Echeverría, 2010, Gómez, Roses and Farias, 2012).

Likewise, university students in the Information and Communication Society are increasingly involved in processes of autonomy in learning, but for this they also need that self-regulated learning be encouraged by active methodologies that integrate social and academic software that facilitate such processes (Schworm and Gruber, 2012).

In this sense, a process of introspective reflection must be carried out on the contributions of the new context assumed by the information society. In this new framework of web 2.0, social networks, convergence in the EHEA, etc., a more flexible and varied perspective of the teaching-learning process is defined, which facilitates the educational response to the diversity of each individual, in such a way that adequate initial teacher training, especially in attitudes and skills, is essential for a quality education for all citizens (Flores, Ari, Inan and Arslan-Ari, 2012).

Finally, in the current European didactic metamorphosis, the establishment of a teaching system that favors the comprehensive training of students oriented to cover the socio-labor demands and lifelong learning under the principle of continuous improvement is prioritized (Arís and Comas, 2011), as the reformulation of methodologies applied in the classrooms, focusing its emphasis on the learning process and giving greater prominence to these students (Salmerón, Rodríguez and Gutiérrez, 2010). And all this requires a faculty that can respond to the rest of the students and institutions.

3.2 Citizenship and the Cyber Students

The population that accesses higher education has a digital experience that is less used and even rejected by the institution. Because traditionally in educational environments, formal and non-formal, the prevailing educational paradigms have been typical of an academic model, characterized by focusing more on teaching rather than learning. Greater importance has been given to training students to memorize certain fragments of knowledge than to enable them to understand and make their knowledge their own (Martínez and Torres, 2013). But the current technological advances (mobile internet, the social web, social networks, cloud computing, etc.) are generating new forms of learning and interaction that have made these academic pedagogical systems obsolete. Tools like blogs, social networks, wikis, collaborative work tools, etc., are generating useful virtual learning spaces of great didactic potential. Currently, the new pedagogical paradigms focus more on the learning processes, working through collaboration, participation, and creativity, leaving behind the final product of knowledge (Aguaded, Pérez and Monescillo, 2010). Despite this, we still find a predominance of training institutions with organizational structures, didactic models and obsolete methodologies, where the textbook remains the king of resources, to the detriment of other more operational, useful, and motivating tools, such as 2.0 resources. In a hypermedia society, there is no room for bookish teaching. The new generation will be the builder of ideas, knowledge, and experiences that circulate through the networks of the future, provided they become familiar with them from their youth.

For Davidson and Goldberg (2009), the digital age has generated unsuspected possibilities for self-learning, the creation of horizontal structures that break with traditional authoritarian schemes, collective credibility, decentralized learning, and network learning, among other aspects. For UNESCO (2012), ICTs can contribute very efficiently to universal access and equality to education and quality learning, to facilitating and improving and the professional development of teachers, as well as to the management and administration management more efficient education system. Area (2012) adds that education, both in formal and non-formal settings, in addition to offering equal access to technology, must train, technologically speaking, the public so that they are more educated, responsible, and critical individuals, since knowledge is a necessary condition for the conscious exercise of individual freedom, and for the complete development of democracy. The two key elements in digital literacy are access and training for critical knowledge.

Digital literacy, therefore, is a question that goes beyond the mere learning of the use of social software tools. The incorporation of ICTs in the classroom must be integrated into educational policies that facilitate access to digital technology and culture for all citizens, so that they "know the technical mechanisms and the forms of communication of different technologies; possess search, selection and analysis skills of the multiple information available on the Web; acquire value criteria that allow them to discriminate and select those products of higher quality and cultural interest; learn to communicate and collaborate in social networks; they are qualified to produce and express themselves through documents of an audiovisual and hypertext nature; know how to bring to light the economic, political and ideological interests that are behind every company and media product; as well as that they become aware of the role of the media and technologies in our daily life. What is at stake is the social model of the information society. Achieving the above goals will mean that this model of future society will be based more on democratic principles and criteria than on merely mercantilist ones" (Area, 2012, p. 5). In other words, it is necessary for citizens to achieve a certain degree of development in digital competences, understood as the ability to use knowledge and skills related to the development of elements and processes. By making use of these skills (knowledge, skills, and aptitudes), the tools and technological resources will be effectively and efficiently used.

Along the same line of arguments, it is obvious that in advanced countries, the socio-technological context is increasing exponentially, and it may serve as an example, the latest report made by the European Commission, from the General Directorate of Information, which regularly conducts opinion surveys to obtain relevant information that shows trends in use and technological resources in both the electronic communications markets and to assess how EU households and citizens benefit from an increasingly competitive and innovative digital environment. Below are some data from the Special Report of the Eurobarometer on the status of Electronic Communications and the Digital Single Market in the European households (Special Eurobarometer 438-European Commission May 2016).[1] This study includes the 28 member states and addresses the following areas:

- Dial-up access via fixed and mobile network
- Internet access through fixed and mobile network
- Means of access to television
- Penetration of communications packages
- Selection criteria when choosing an Internet provider
- Ease of package comparison and ease of change of provider
- Knowledge of the European single emergency service number 112

Of the results offered by the report, we consider the referrals to dial-up and Internet to be interesting for this work. In relation to telephone access, this is almost universal, with 98% penetration. In 93% of the European households, there is access to mobile telephony through at least one device. In 65% of these households, there is access to fixed and mobile telephony, with 33% of the households where there is only access to mobile telephony. In this sense, mobile telephony continues to increase persistently, as the proportion of mobile lines, which has increased by 18% since the end of 2005. In Spain, 71% of the households have access to fixed telephony and 92% mobile telephony. Figure 3.1 shows the evolution of the different modes of dial in Europe from 2005 to 2015.

According to this study, mobile telephony is by far the most important communication service in the daily life of the people surveyed (74%), followed by mobile Internet (34%) and fixed telephony (32%). In addition, Internet connections (52%) and online communication services (46%) are the most important services in the daily life of around half of the European citizenship.

In terms of Internet access (Figure 3.2), a little more than two thirds (67%) of the European households have Internet access in their homes, but penetration in the different member states varies widely: 41% in Italy or 57% of Spain to 96% in the Netherlands. At the EU level, Internet penetration in households has increased by 10 points since 2009. However, it is worth noting the decrease in Internet access in households observed in ten member states since 2014. This relatively small decrease in most of the countries can be explained in part by the general increase in mobile Internet access.

[1] European Commission. Special Eurobarometer 438. E-Communications and the Digital Single Market Report (May 2016). URL: http://bit.ly/2oy4yBp

Figure 3.1 Evolution of telephone access in European homes (%).
Source: EC. E-Communications and the Digital Single Market Report 2016. (Own preparation)

In this sense, access to mobile Internet has increased considerably, as in 69% of households in the EU have at least one of its members with Internet access on their mobile. Access to mobile Internet has also increased in all member states, and the proportion of households with mobile Internet access ranges from 91% in Denmark (16 points since 2014), 70% in Spain, and 64% in Greece (35 points).

Finally, as a point of interest, almost all households in the EU have access to television (96%), digital terrestrial television (38%), satellite (24%), and digital cable (20%). most common reception means. At the country level, access to digital terrestrial television varies from 90% in Spain to 5% in Slovenia.

Another report of interest is the one that is periodically published in the Global Web Index (GWI) website,[2] specialized in usage trends about the Internet, social networks, digital content, and electronic commerce. In the last study published in the first quarter of 2017, some very interesting indicators for collective reflection are shown.

[2]Global Web Index. URL: http://globalwebindex.net/

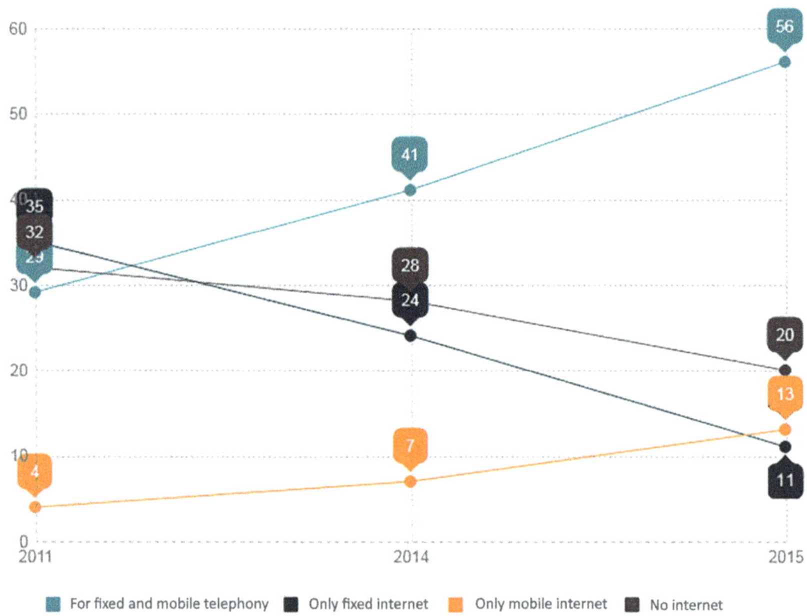

Figure 3.2 Evolution of the modalities of Internet access in European households (2011–2015).
Source: EC. E-Communications and the Digital Single Market Report 2016. (Own preparation)

According to this report, more than 50% of the Internet users spend more than 3 hours a day connected, while more than 90% have a smartphone or mobile. The connection to information, news, and events around the world is instantaneous. The average user, when connected, spends on average 47% of the time browsing different services, 24% of the time watching online television, 13% using streaming music or radio services, 10% consulting the daily press, and 6 % of the time in video games.

Regarding the use of devices, from the information represented in Figure 3.3, it can be inferred that, in the medium or long term, probably more than 50% of the consumers of multiple devices (Smartphone, Tablet, PC, Laptop, etc.) will soon choose the mobile as your favorite device. But this is where demographic data becomes the key, as there are already certain audiences where this is the case. In this sense, there is a clear correlation between age and the importance of the device: for example, in people of age between 16–24 years, the smartphone has exceeded the 50% rate in terms of preference, and the range of 25–34 years is very close to that number.

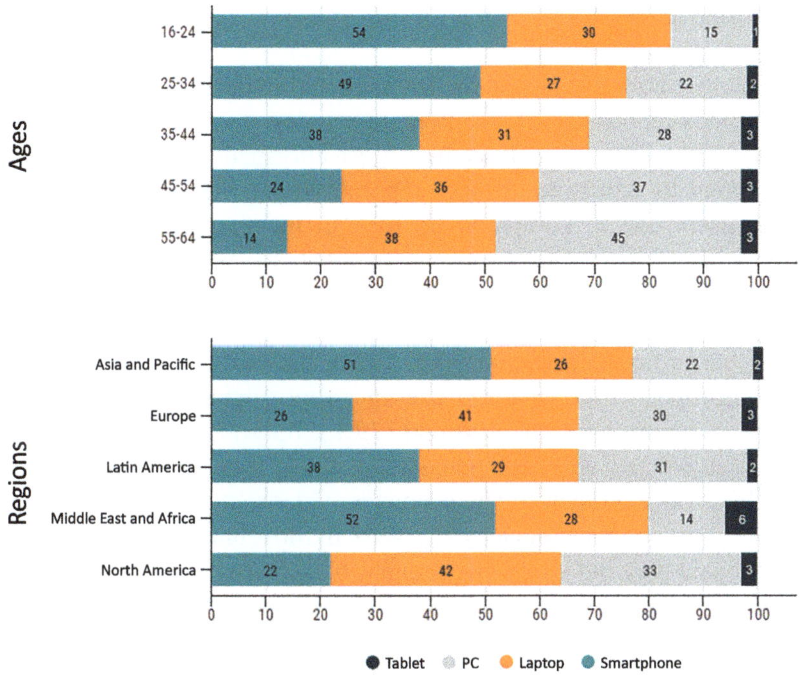

Figure 3.3 Importance of devices according to users by age and region.
Source: GWI. Trends 17. The trends to watch in 2017. (Own preparation)

From 55–64 years old, this preference drops profusely to reach only 14% in terms of preference.

Something similar happens with these data by regions. In Europe and North America, regions where "traditional" devices (PC and Laptop) were the common access tool, mobile devices struggle to overcome 25% when it comes to the importance of devices. On the contrary, more than 50% of users of multiple devices from the Asia-Pacific and the Middle East/Africa have a preference for the mobile device, as they consider it to be the most important device. There is also the fact that precisely in these regions, there are more deficient infrastructures for access to the Internet through fixed broadband.

In convergence with the previous reflections, in a report of the European Commission published in June 2016 about the use of online platforms,[3] it was concluded that almost nine out of 10 Internet users use search engines at least

[3]European Commission. Special Report 447. Online platforms. June 2016: URL: http://bit.ly/2oGO76i

once a week, and therefore the search for information is an important service in terms of Internet use.

In relation to search engines on the Internet, and returning to the previous Global Web Index report, Google maintains a privileged position that borders on monopoly, consolidating itself as the largest information controller in the world, beginning to dominate all access points to Internet. Approximately 85% of the users use it, compared to 76% in the first report published five years earlier. Even in countries where there are very powerful local alternatives, as in Russia with Yandex, it is only 8 points (82% versus 90%).

Another most used service offered on the Internet is social networking. In this sense, according to Special Report 447 of the European Commission, 60% of the European Internet users use an online social network at least once a week. Countries such as Portugal (79%), Italy (72%), and Bulgaria (71%) stand out, where a very high percentage of people surveyed use a social network at least once a week to share images, videos, etc. On the contrary, only 42% of the respondents in Germany, 49% in Slovenia, or 51% in France and Denmark said they did so. In our country, 67% (above 60% of the EU as a whole) of the Internet users surveyed use a social network at least once a week. Most of these people also use e-commerce occasionally, although their use is less frequent compared to the search engines or social networks: only 30% use them at least once a week (Figure 3.4).

Regarding the use of personal information by these platforms, almost two thirds of the people surveyed (64%) are aware that the Internet experience may be different for each person based on their previous online activities. In this sense, at least half are concerned about the data collected from them when they are online and more than half (55%) agree that online platforms should be regulated by public authorities to limit the extent of arbitrary and commercial control. They show different search results to users based on the data collected about them. Only a minority (30%) is comfortable with the search engine using information about their online activity and personal data to adapt ads or content to their interests. A little over a quarter are comfortable with this fact when it comes to online markets (27%) or social networks (26%). However, paradoxically only one third of the people surveyed said they read the terms and conditions of the web services, and only about one in five (19%) said to take them into account when using the online platform in question.

Focusing on social networks and messaging as the most used services on the Internet, the Statista web, specialist in online marketing, offers interesting data regarding the use of social networks. Currently, it is estimated that there

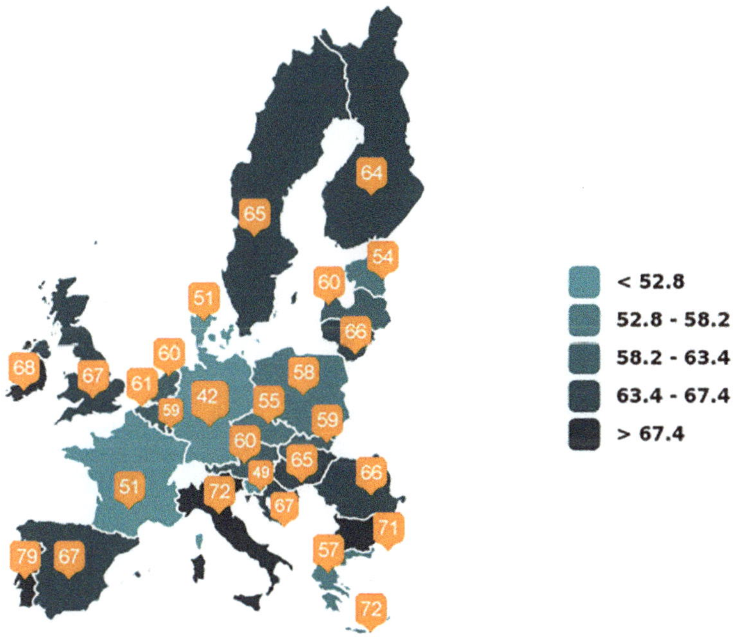

Figure 3.4 Percentage of users who use social networks at least once a week (per country). *Source:* European Commission. Special Report. Online platforms. June 2016. (Own preparation)

are just over 2.500 million users of social networks and this figure, constantly increasing since 2010 (with 970 million users), is expected to continue to grow as the use of mobile devices and mobile social networks win more and more strength.

The main social networks are usually available in several languages and allow users to connect with family, friends, or other people regardless of the geographical, political, or economic boundaries. The most popular social networks tend to have a high number of user accounts with a strong commitment on their part. The use of the social network by consumers is very diverse: platforms such as Facebook or Google+ are very focused on exchanges between friends and family and are constantly facilitating interaction through sharing photos or social games and statuses. Other social networks such as Tumblr or Twitter focus on faster communication and are called microblogs. Some social networks focus on the community; Others highlight and display the content generated by the user.

Due to a constant presence in the lives of its users, social networks have a strong social impact. The confusion between offline and virtual life, as well

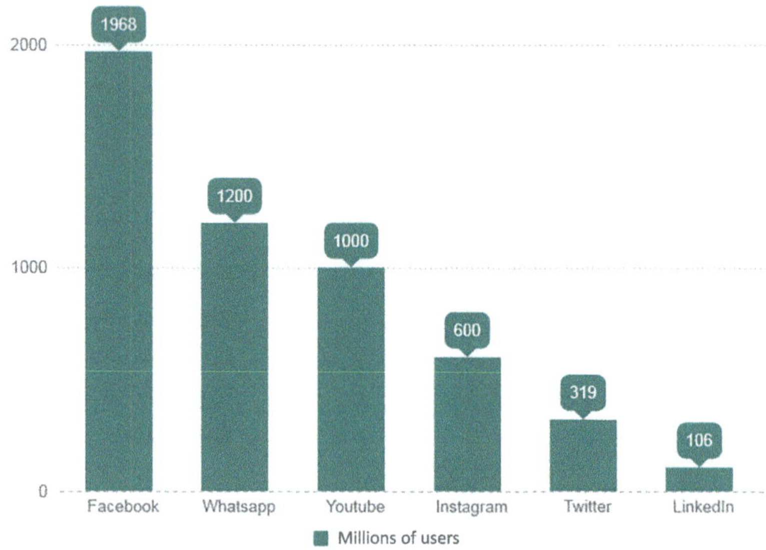

Figure 3.5 Active profiles in the main social networks. April 2017.
Source: Statista. The Statistics Portal.[4] (Own Elaboration)

as the concept of digital identity and online social interactions, are some of the aspects that have emerged in recent discussions.

In terms of the number of users per social network (Figure 3.5), Facebook is the leading social network in the world in terms of popularity. It was the first to overcome the 1,000 million registered accounts and is on track to reach 2,000 million (1970 million). In popularity, it is followed by the well-known messaging service WhatsApp (owned by Facebook), which has some 1.2 billion active accounts, and YouTube (owned by Google) that has at least 1,000 million active users. Instagram, the popular photo-sharing application, also owned by Facebook, ranks seventh with more than 600 million active monthly accounts. Meanwhile, the well-known Twitter microblogging service slightly exceeds 319 million users.

Following this line, in the 8th Observatory of Social Networks of December 2016[5] (TCAnalysis, 2016), there is a penetration of users in social networks, which has been stable since 2011, being 91% in 2016. Facebook remains as the social network par excellence with 88% of the Internet users with an active account. For its part, Instagram succeeds in unseating Twitter

[4]Statista. The Statistics Portal URL: http://bit.ly/2pfw9sc
[5]VIII Observatory of Social Networks. December 2016. URL: http://bit.ly/2pczFTa

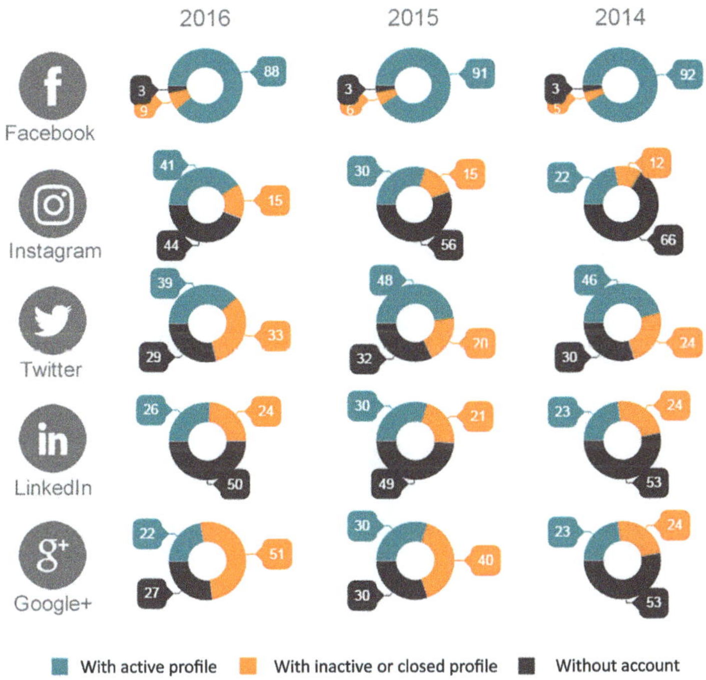

Figure 3.6 Report on social networks. Evolution in the profile status.
Source: VIII Observatory of Social Networks (December 2016). (Own elaboration)

from the second position, consolidating its growth with a penetration of 41% of the Internet users, while Twitter continues to decline and stagnates at 39% penetration.[6] LinkedIn remains at levels very similar to 2015, both in penetration and in use and Google+ returns to 2014 levels, with a 22% penetration, after a slight rise in 2015, obtaining in the identical way, as it is verified in the same, and it continues to increase (Figure 3.6).

Focusing on our country, the National Observatory of Telecommunications and the Information Society (ONTSI, 2017), has recently published its report on the socio-demographic profile of Internet users from the analysis of data offered by the INE with respect to 2016 (Table 3.1). The number of people of at least 10 years of age who have accessed the Internet once exceeded 31 million and a half in 2016. Of these, more than 27 million people

[6]Note: Only data referring to global social networks are being taken into account, without taking into account local social networks such as QQ, QZone or Seina Weibo, which are very popular in China and have hundreds of millions of active users.

Table 3.1 Use of the Internet by sociodemographic segments

		Internet Users			Netizens Last Month			Internet Access Weekly		
		2014	2015	2016	2014	2015	2016	2014	2015	2016
Sex	Man	80,1	82.8	84.6	76.2	79.2	81.6	73.4	76.5	78.3
	Woman	77.1	79.2	80.8	72.5	75,7	77.7	69.0	73.0	74.6
Age	From 16 to 24 years old	98.5	99.1	99.2	97.5	98.2	98.2	96.2	96.8	96.8
	From 25 to 34 years old	95.4	96.3	96.8	91.9	93.1	95.3	89.9	91,7	93.8
	From 35 to 44 years old	92.3	93.8	94.9	88.1	90.5	92.5	84.2	87.7	89.2
	From 45 to 54 years old	81.8	84.7	87.7	75.6	80.2	83.7	70.6	76.0	79.1
	From 55 to 64 years	58.4	64.8	68.0	52.9	59.8	63.3	50,1	56.8	59.1
	From 65 to 74 years old	28.8	33.9	38,1	25.1	30.5	33.5	22.8	28.0	30.7
Employment situation	Employed as an employee	92.3	93.1	94.7	89,0	90.6	92.4	86.2	88.5	89.6
	Occupied on your own	86.1	88.6	91.6	81.6	85.3	90.1	77.6	82.5	85.6
	Active stopped	81.3	85.4	83.6	74.3	79.5	78.6	70.2	74.9	74.3
	Student	99.1	98.2	99.5	98.7	97.9	99.2	97.8	97.2	98.0
	Housework	42.4	45.4	48.4	37,1	40.5	44.4	32.5	36.3	40.5
	Pensioner	40.3	44.1	48.7	35.6	39.9	43.5	32.9	37.2	40.2
	Other employment situation	62.6	60.5	61.3	57.2	52.2	55.4	55,7	51,4	52.1
Finished studies	Illiterate and incomplete primary	13.4	14.1	16.7	9.7	11,1	15.4	8.4	8.2	12.4
	Primary education	40.5	42.6	44.2	34,7	36.9	39.0	30.5	33.3	34.8
	1st stage of ed. high school	76.8	79.7	84.4	70.2	74.5	79.8	65.5	70.1	74.0

Table 3.1 Continued

		Internet Users			Netizens Last Month			Internet Access Weekly		
		2014	2015	2016	2014	2015	2016	2014	2015	2016
	2nd stage of ed. high school	92.9	94.4	94.2	89,1	90.8	91.8	86.3	88.8	89.2
	FP grade higher	95.3	97.3	97.6	91.8	94.3	95.8	89.5	91.6	93.7
	Higher university Ed.	98.5	98.6	98.8	97.2	97.7	97.3	95.6	96.9	96.6
	Can not code	33.7	38,1	68.1	20.3	38,1	68.1	20.3	38,1	63.7
Habitat size	Capitals > 500 thousand hab.	83.8	87.3	88.9	80.5	84.6	86.1	78.8	83.6	83.9
	Capitals < 500 thousand hab.	82.2	84.9	85.4	79.0	81.3	82.7	75.6	78.6	79.9
	Municipalities > 100 thousand inhabitants	77.1	81.3	81.6	73.0	78.5	78.6	69.3	76.1	76.0
	Municip. 50 to 100 thousand people	81.6	82,4	84.7	77.8	79.0	81.6	74.6	75,7	78.2
	Municipalities 20 to 50 thousand inhabitants	77.6	78.9	80.8	72.4	75,7	78.1	68.9	73.5	74,7
	Municipalities 10 to 20 thousand inhabitants	77.1	80.8	82.2	72,1	77.1	78.4	70.1	73.6	74.6
	Municipalities < 10 thousand hab.	72.9	74.2	76.5	68.1	69.3	73.0	64.1	65.7	69.2
Net income per household	Less than 900 euros	61.4	65.0	63.3	54.1	58.4	57.6	49.5	53.9	53,7
	From 901 to 1,600 euros	76.2	75.8	79.0	71.8	71.6	75.3	67.9	68.7	71.3
	From 1,601 to 2,500 euros	88.5	89.5	89,8	86.0	87.3	87.6	83.5	85.2	85.3

(*Continued*)

Table 3.1 Continued

		Internet users			Netizens last month			Internet access weekly		
		2014	2015	2016	2014	2015	2016	2014	2015	2016
	From 2,501 to 3,000 euros	94.5	96.6	96.7	91.8	95.1	94.9	90.4	92.9	93.6
	More than 3,000 euros	97.9	96.1	97.8	97.0	95.6	97.6	95.5	95.3	96.0
	Ns/Nc	78.5	82.9	84.7	74.4	79.8	82,0	72,1	77.5	78.6
Nationality	Spanish	77.7	80,1	82,0	73.8	76.9	79.0	70.9	74.5	75.9
	Foreign	82.8	86.4	86.5	76.4	79.7	82.5	71.6	74.8	79.2
	Spanish and other	93.5	90.5	90.5	90.0	87.5	88.9	83.8	87.5	86.0
	Any	79.2	100.0	100.0	79.2	100.0	100.0	79.2	81.9	100.0
Civil status	Single	91.5	—	93.7	88.6	—	91.6	86.8	—	89.4
	Married	73.0	—	77.8	68.2	—	74.5	64.2	—	70.6
	Widower	32.0	—	44.0	25.6	—	40.0	24.5	—	37.9
	Separated	74.3	—	76.8	68.8	—	68.0	66.5	—	64.1
	Divorced	85.0	—	88.9	79.3	—	84.4	74.2	—	81.8

Source: Socio-demographic profile of Internet users. INE 2016 (ONTSI, 2017)[7]. (Own elaboration)

[7] Socio-demographic profile of Internet users. INE data analysis 2016 (ONTSI, 2017). URL: http://bit.ly/2oCve1C

aged between 16 and 27 years of age were connected in the last month and almost 23 million connected to the Internet daily.

In general data, the Internet user population between 16 and 74 years of age does not show substantial differences in terms of sex, although the number of men (84.6%) is slightly higher than that of women (80.8%). The differences are appreciable when the variable age is taken into account, although we could speak of two differentiated groups. On the one hand, it includes the age groups between 16 to 24, 25 to 34 and 35 to 44 years old, which have similar data regarding access (96.8%, 93.8% and 89%, 2% respectively in relation to weekly access). From these ages, access to the Internet begins to decrease progressively, being 79.1% for users 45–54 years of age, 59.1% for Internet users aging between 55 and 64 years, and 30.7% for Internet users aging between 65 and 74 years old.

Taking into account the variable monthly income, 63.3% of the people who reside in households with monthly rent of less than 900 € were connected to the Internet, while that percentage rises to 97.8% in the case of a greater income (€ 3,000).

As can be seen, there is a clear boom in Internet access in our country. Likewise, there is also a boom in the consumption of digital content through the Network as reflected in the Annual Report of the ICT Sector and the Contents in Spain 2016[8] (ONTSI, 2016), in which the main data of the Digital Content Industry of our country are presented. This study provides relevant data in relation to the consumption habits of content by citizens as well as trends in each of the sectors studied. Some of the main data of the report in relation to its importance as an economic sector are:

- The digital content industry in Spain invoiced a total of 15,467 million euros in 2015, 13.2% more than in the previous year, being also the second consecutive year in which there is an increase after several negative years.
- The activities that contributed most to this increase were programming and broadcasting of radio and television, with a turnover in 2015 of 4,222 million euros (11.5% more than in 2014); cinematographic activities, of video and of television programs, that invoiced 3.001 million euros (16.7% more with respect to 2014), publication of books, newspapers and other activities of publication with 5.702 million (with a 7.7%

[8]Annual Report of the ICT Sector and of the Contents in Spain 2016. URL: http://bit.ly/2nPVDJp

year-on-year growth). These three activities represent 83.6% of the total billing of the content sector.
- In relative terms, the branches that grew the most were those of other information services, sound recording activities, and music and videogame edition (46.6%, 32.5%, and 24% respectively).
- The turnover of the sale of software for video games in Spain amounted to 511 million euros, continuing this way, after a decrease in 2011, with an upward trend and approaching 2010 data in which 575 million were invoiced.
- Digital music experienced an increase of 10% in its turnover, reaching 319 million euros.
- Advertising investment in digital media was 1,289 million euros in 2015.
- 91.5% of the Spanish population consumes digital content through the Internet or through an electronic device not connected to the Internet.

Table 3.2 summarizes the data provided in this annual report shows.

Based on the data presented, it is more than evident that we are witnessing a technological development never imagined. These advances generate innumerable doubts, uncertainties, and fears, and given that the educational field, on which this work focuses, is one of the most influential, we believe it is necessary to infer some pedagogical questions linked to ICT (López-Meneses, 2012):

Table 3.2 Number of businesses in the Content sector (Millions of Euros)

Content categories	2010	2011	2012	2013	2014	2015
Publications of books, newspapers and other publication activities	7.175	6,788	5,993	5.261	5,292	5,702
Motion picture, video, and television program activities	3,284	3,296	3.018	2,721	2,572	3.001
Sound recording and music editing activities	370	340	295	272	241	319
Other information services	293	263	243	230	289	423
Programming activities and broadcasting of radio and television	4,421	4.124	3,761	3.613	3,786	4.222
Video game	575	499	428	314	412	511
Online advertising	799	899	881	960	1.066	1,289
Total	16,918	16,208	14,618	13,372	13,658	15,467

Source: Annual Report of the ICT Sector and Content in Spain 2016. (ONTSI, 2016)

- What can happen when so many people communicate, if we remember that educational processes are eminently communicative processes?
- Will social networks be the ideal virtual space for freedom of expression?
- Will the virtual courses be designed according to psycho-pedagogical criteria, or rather, only and exclusively from economic interests and marketing fads?
- Can it be true what Wolton (2004) points out that the world has become a global village on a technical level and is not on a social and cultural level?
- Will it be emotional usability (Emotional Design)[9], a significant criterion for the design and development of digital communication environments?
- Is there a true inter-culturalism of races, religions, and cultures on the Web or, rather, a monopoly of certain predominant cultures?
- Will it encourage personal and interpersonal enrichment for the humanistic formation of the individual or, on the contrary, will it generate a collective of anonymous masses?
- Can tomorrow's information reside about our human genome, economic status, confidential data and privacy in the hands of Hackers and Crackers? And what can happen in relation to information on national defense or international financial institutions?
- Is it possible that these technologies are promoting a kind of virtual sociability to the detriment of human sociability, understood as the interpersonal relationship in thc samc time and space?
- How will the revolution that the progressive transformation of the Web towards the so-called Web 2.0 and the implementation of MOOCs and augmented reality affect the youngest?
- Will it be Web 2.0. one of the sources of knowledge or the genesis of the chaos of information?
- And with regard to the processes of training and ongoing research, will it lead us to the development of professional autonomy and development

[9]Kansei engineering incorporates emotion and affection in the design process. Kansei is how a user perceives a product (user experience). Also, it is defined as Sensorial Engineering or Emotional Usability. By means of this technique, those attributes of a design are detected that allow obtaining certain subjective responses from people and design based on the pursuit of those responses. For this method objects are used that allow obtaining extreme responses: pleasant-unpleasant, attractive-ugly, easy to use-complicated, simple-complex. http://www.grancomo.com/glosario.php?x=K

or will we become more dependent on new technological resources, functioning as true prostheses?

- With such a rapid pace of change, to what extent are we able to face this high level of instability and uncertainty?
- Is it true that the Internet is a worldwide network that can be accessed by anyone from anywhere in the world? And with any disability?
- How to locate in an efficient way all this acerbic and accumulation of information, news, articles and scientific comments?
- If most of our students have been growing in passive and receptive learning environments, and most prefer that the universities be the extension of the kindergarten, or, of the institute, where they tell them what they have to do, how to motivate them in more active and participative non-physical training environments?
- Will we be more human and happy walking towards the formation of Homo Media?

4

MOOC: Strengths and Weaknesses

The nature of information and communication has been characterized in recent years by a formative commitment based on activities, courses, and proposals based on technology-mediated teaching-learning processes (Castaño-Muñoz, Duart and Teresa, 2015; Estévez and García 2015, Roig-Vila, Mondéjar and Lorenzo-Lledó, 2016, Colorado, Marín-Díaz and Zavala, 2016; Ponce, Pagán-Maldonado, y Gómez Galán, 2018). Under this technological scenario, the so-called MOOC shines its own light.

The acronym "MOOC" literally translates as Massive Open Online Course (Rheingold, 2013). In other words, it is clearly defined by its openness, by locating the information and the relationship between the different educational actors on the Internet ("online"), and by the fact that the size of the educational community involved in a course of these characteristics can easily exceed thousands of people ("massive"). MOOCs displace (some would say "exceed") the hierarchical relationship between teacher and students, so that the learning process is shared (hence the references in the MOOC literature to the idea of a "distributed responsibility" in learning), and students become, also, a generator of content and connections between different aspects of the course. Emphasis is placed, in MOOCs, on the use of social networks (Facebook, Twitter, etc.) with which it is intended to consolidate these learning communities. In addition to social networks, members of the learning community can take advantage of content aggregation (RSS, for example) to share information, thematic or tangential materials, and learning strategies (Méndez-García, 2013). In this sense, the new university training scenarios are moving towards a new model of mass education, open and free through a methodology based on video-simulation and collaborative work of the student (López-Meneses and Vázquez-Cano, 2017; Gómez Galán, 2018).

In this respect, the MOOCs have captured a worldwide interest due to their great potential to offer free training (Rabanal, 2017), of quality and accessible to anyone regardless of their country of origin, their previous

Figure 4.1 Some characteristics of MOOCs. (Own elaboration)

training, and without the need of paying for your registration (Liyanagunawardena et al., 2013). Among the fundamental characteristics of MOOCs (Figure 4.1), the following stand out (McAuley, S tewart, and Cormier, 2010, Siemens, 2013):

- Free access;
- No limit on the number of participants;
- Absence of certification for free participants,
- Instructional design based on the audiovisual with support of written texts;
- Collaborative and participatory methodology of the student with minimal intervention of the teaching staff.

According to Castaño and Cabero (2013, p. 89), MOOCs are also characterized, in addition, to their gratuity:

- It is an educational resource that has a certain similarity with a class, with a classroom.
- With starting and ending dates;

- It possesses evaluation mechanisms;
- It is online.

Currently, MOOCs are classes taught through technological platforms that enable access to the teaching-learning process thousands of users (Ramírez-Fernández, 2014). In summary, as indicated by Aguaded and Medina (2015), the MOOC movement (Massive and Open Online Courses, in Spanish COMA) arises from an innovation process in the field of open knowledge formation, guided by the principles of mass dissemination, and free of the contents and intermediated by online, interactive and collaborative application models.

On the other hand, the phenomenon of MOOCs is relatively recent (Graham and Fredenberg, 2015). In 2008, the term "MOOC" was introduced in Canada by Dave Cormier (2008), who coined the acronym to designate an online course conducted by George Siemens and Stephen Downes. The course entitled "Connectivism and Connective Knowledge" was conducted by 25 people who paid their tuition and obtained their degree, but it was followed free of charge and without accreditation by 2300 students and the general public through the Internet (Downes, 2012a, Daniel, 2012).

After this experience, the second successful attempt to export this idea materialized in the summer of 2011 in which the Stanford University offered a course on "artificial intelligence" online in which 58,000 students enrolled. One of the people involved in the project was Sebastian Thrun, later founder of the MOOC platform "Udacity" (https://www.udacity.com) that provides support to important universities for the development of open training (Meyer, 2012).

The Massachusetts Institute of Technology (MIT), meanwhile, initially created MITx for the design of this type of courses, but this platform eventually evolved into a joint platform of the Harvard University, the University of California-Berkley, and the MIT itself with the name of EDx (https://www.edx.org). Coursera (https://www.coursera.org) (Lewin, 2012; DeSantis, 2012) is the platform that has most developed these initiatives and what is being signaled as the standard-bearer in pedagogical design. In Figure 4.2, the evolution is described in the genesis of the MOOC courses.

The New York Times (Figure 4.3), named 2012 as "The Year of the MOOCs" publishing an article that highlighted the great impact that the MOOC courses were having and predicting that these would become a tsunami that would sweep the traditional universities. (Pappano, 2012).

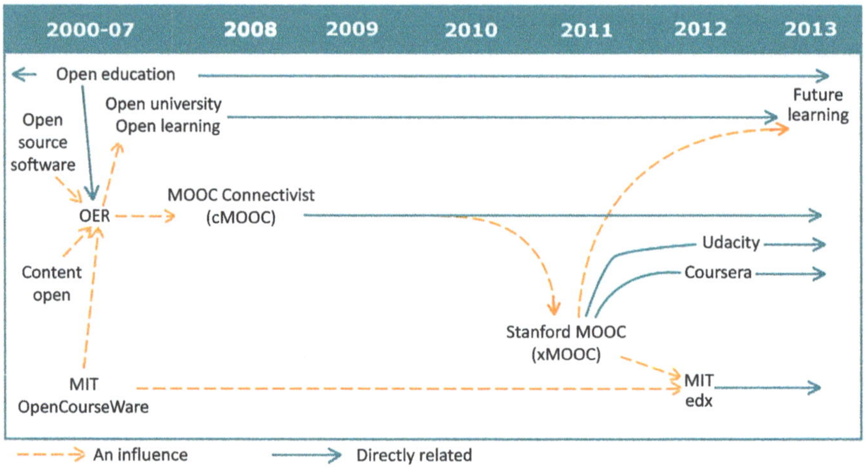

Figure 4.2 Timeline of the genesis of MOOCs and open formation. Adapted from White Paper "MOOC and Open Education: Implications for Higher Education".

Figure 4.3 The NYT (2012 MOOC Year).
Fuente: http://nyti.ms/ShTBdq

Since then, this phenomenon has placed online, massive and open training in the media showcase. Although there are authors who point out that a turning point has been reached in Gartner's popularity curve (Hype Cycle), and that the novelty effect tends to subside (Urcola Carrera and Azkue Irigoyen, 2016), it is enough with a consultation through the Google Trends media trends tool to know the boom caused by MOOCs since then and how this boom is even more in force, if possible. Figure 4.4 graphically shows how the media interest aroused began a clear rise in 2012 which, with some fluctuations remains high, reaching the maximum score (100) on several occasions since 2012 and being the last in March 2017.

The Horizon report, led by the New Media Consortium and Educause, provides a prospective study of the use of technologies and educational trends in the future of different countries. In its ninth edition (Johnson et al., 2013), the incidence of MOOCs in the current educational panorama stands out. Likewise, the Ibero-American edition oriented to Higher Education, joint initiative of the "eLearn Center" of the UOC and the New Media Consortium, indicates that the "massive open courses" will be implemented in our institutions of Higher Education in a horizon of four to five years (Durall et al., 2012).

Different authors consider that this new type of format actively promotes self-organization, connectivity, diversity and decentralized control of teaching-learning processes (Dewaard et al., 2011, Baggaley, 2011, Vázquez-Cano and Sevillano, 2013).

In this sense, MOOCs are being considered by many researchers as a revolution that is beginning to affect the traditional structure of university and training organization (Boxall, 2012; Weissmann, 2012) and whose development in a very close horizon is exciting, disturbing and completely unpredictable (Lewin, 2012).

The repercussion of the MOOC movement is significant not only in the formative and academic world but also in its representation in blogs, news and reports generated in recent years. The search engine of Google so testifies with more than 12 million entries and a curve that has increased substantially as can be seen in Figure 4.4.

MOOCs as indicated by Gértrudix, Rajas and Álvarez (2017) are being widely treated in the academic literature in a way that goes from

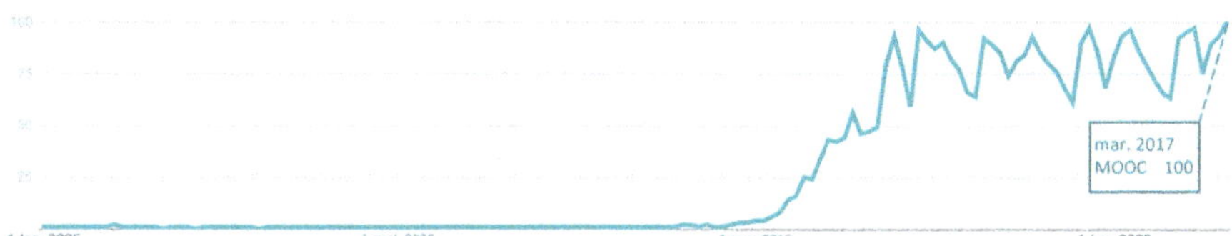

Figure 4.4 Media relevance of the term MOOC in the Google search Enghien.
Source: Google Trends. Search made in April 2017. URL: http://bit.ly/2odNttE

bibliometric analysis that measure the representation of the concept in the scientific literature and, therefore, its interest as an object of study (López-Meneses, Vázquez-Cano and Román-Graván, 2015; Zancanaro and Domingues, 2017), the institutional policies that stimulate them (Hollands and Tirthali, 2014), in the academic context due to the disruptive innovation they bring to the educational system (Zancan a o and Domingues, 2017), or the examination of its pedagogical quality (Roig-Vila, Mengual-Andrés and Suárez-Guerrero, 2014, Aguaded and Medina-Salguero, 2015), among other areas.

As it could not be otherwise, the universe of MOOCs is the object of a continuous didactic and formative reflection among different authors (Daniel, 2012, Vázquez-Cano et al., 2013, Zapata-Ros, 2013, Gómez Galán, 2014b y 2018; Ramírez-Fernández, Salmerón-Silvera and López-Meneses, 2016; Gómez Galán and Pérez Parras, 2017, García, Fidalgo, and Sein, 2018). At the same time, different studies carry out systematic studies on the research carried out in MOOC between 2008 and 2013 (Liyanagunawardena et al., 2013, Castaño, 2013, Karsenti, 2013). And studies (biennium 2013–14) that show an upward trend in the volume of publications and the predilection for journal articles and, to a lesser extent, presentations at congresses. The most researched topics have been those related to assessing pedagogical strategies and, especially, motivation and involvement of students (Sangrá et al., 2015). And the number of research and universities that are integrated into this socio-technological phenomenon continues to increase exponentially (Vázquez-Cano, López-Meneses and Barroso, 2015). Also, it is worth highlighting the studies of López-Meneses, Vázquez-Cano, and Román (2015) concerning the bibliometric study comprised between the years 2010–2013, highlighting that the universities and countries, which until now, are having a greater scientific impact are United States, Australia, Canada, United Kingdom, and Spain. And, on the other hand, it is worth highlighting the bibliometric study of Aguaded, Vázquez-Cano and López-Meneses (2016), on the impact of the MOOC movement in the Spanish scientific community that shows that the impact of Spanish scientific production in book format and articles in prestigious internal databases (Wos-SSCI/Scopus) is still low, although the national impact according to ANEP/FECYT and In-Recs categorization is moderately high. And, recently, the bibliometric research of Mengual-Andrés, Vázquez-Cano, and López-Meneses (2017) that analyzes the scientific productivity of the MOOC phenomenon from the analysis of 752 publications indexed in the SCOPUS database in the period 2012–2016. It shows outstanding scientific productivity and consolidates the MOOC phenomenon as a research area and has its starting point in 2012.

This question highlights the validity of the MOOC theme and the confrontation that occurs between researchers in relation to their possibilities and innovative characteristics (Gómez Galán and Pérez Parras, 2016a). Finally, it is noted that the fields of Social Sciences, in particular, and Computer Science are the most active in the research and revitalization of knowledge with respect to MOOC, having produced 31.2% of total scientific productivity of the phenomenon only in 2015, plus 20% occurred in 2014.

Currently, the movement in Spain has had a great impact, more so than in the rest of Europe. For example, the Polytechnic University of Valencia and the UNED have developed their own platform and, at the same time, they are on other aggregator platforms, but most of the universities have their courses mainly on the Miriadax platform.

According to the European Commission, Spain leads the offer of MOOC courses in the European Union with 481 courses (Figure 4.5) together with the United Kingdom with 435 courses. The rest of the countries of the EU

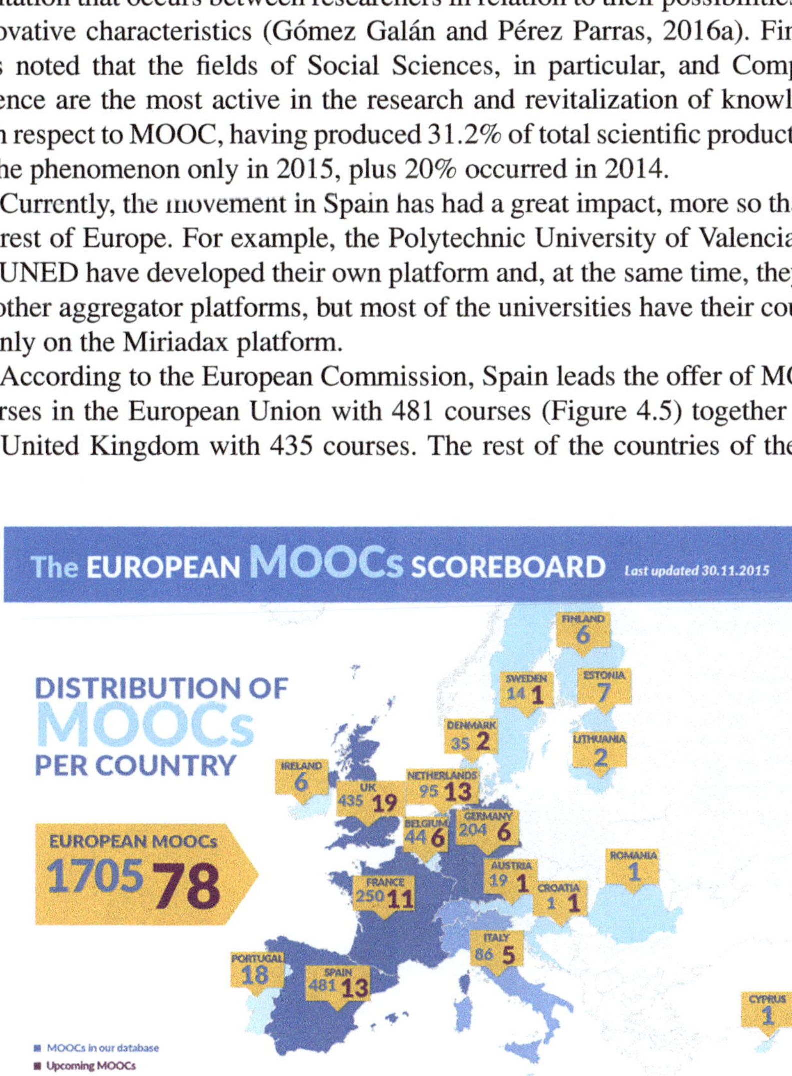

Figure 4.5 Offer of MOOC courses in the European Union (2015).
Source: Open Education Europe http://openeducationeuropa.eu/

are very far apart in this regard, highlighting France (with 250 courses) and Germany (with 204 courses).

As regards the distribution of MOOC courses in the European Union by knowledge area, most of them focus on subjects related to Science and Technology together with the Social Sciences. In a second block, and to a lesser extent, they deal with subjects related to Humanities, Economics, or Applied Sciences. The compendium of subjects closes with the Mathematics and Statistics, the Natural Sciences and in a very superficial way the Arts (Figure 4.6).

As mentioned, in line with Oliver et al. (2014), it can be inferred that Spain has been located in a very short time, and surprisingly, in the leading group of countries that are generating more activity around mass open online courses or MOOCs.

Certainly, in this socio-educational revolution, Spain is adopting a very relevant role in the European and world context. In fact, it has been the leading European country in offering MOOC courses from 2013 to 2016, with hundreds of courses offered. Likewise, the demand for these courses positions Spain among the five countries with the most students in this training modality, with only countries like the United States, the United Kingdom, Canada, and Brazil ahead of the world (Aguaded, Vázquez-Cano, and López-Meneses, 2016). Finally, as indicated by different authors (López-Meneses,

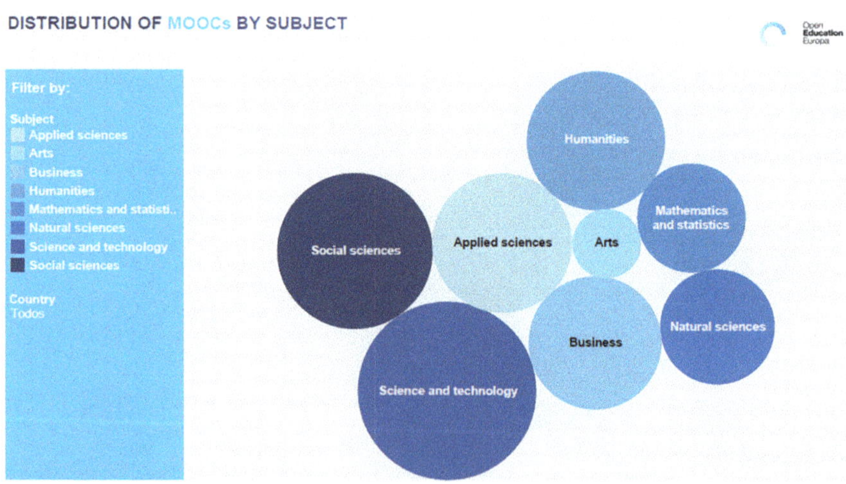

Figure 4.6 Distribution of MOOCs in the European Union (2016).
Source: Open Education Europe http://openeducationeuropa.eu

Vázquez-Cano, and Román, 2015) there is an ascending increase of scientific articles related to this topic worldwide since 2013 up to the present.

Also, as explained in Vázquez-Cano, López-Meneses, Méndez, Suárez, Martín-Padilla et al., (2016) and Martín-Padilla (2015), these massive open courses in network can be the new deposits of cognitive reflection and recreation, the new communication and innovation habitats in the university digital ecosystems and the seed of new massive learning scenarios.

Ultimately, the philosophy of this training modality may imply a democratization of higher education (Finkle and Masters, 2014; Dillahunt, Wang and Teasley, 2015) and despite its growing popularity and prominence, the most promising value of MOOCs is not that it derives from what they are, but from what they can become, that is, from the positive derivatives that are beginning to emerge and that derives from the flexible and open character of the learning they advocate (Yuan and Powell, 2013).

4.1 Taxonomy of MOOCs: xMOOX vs cMOOC

Nowadays, the massive arrival of mass online and open courses has been considered in scientific and informative literature as a revolution with a great potential in the educational and training world (Bouchard, 2011; Aguaded, Vázquez-Cano and Sevillano, 2013, Ramírez-Fernández, Salmerón and López-Meneses, 2015).

Massive courses, open and online, are presented under a diversity of organization and design, which not only imply different visions about what the training process should be, but also regarding what the students should do, the forms in which they should be evaluated, what they should do in them, and the ways to design the contents (Cabero, Llorente and Vázquez, 2014).

When designing MOOC courses, sometimes you rely on instructive pedagogies, in which the teaching follows a model of unidirectional transmission. They either follow a behavioral, cognitivist, or constructivist model, or a study guide or activity proposals, which facilitate the information that students use during their learning. In most of these proposals, students are not treated as a source of knowledge but as a receiver or, at best, as a builder (Dron and Ostashewski, 2015).

The most generalized typology of MOOC courses is the difference between xMOOC and cMOOC (Downes, 2012b, Hill, 2012, Siemens, 2012, Karsenti, 2013, Vázquez-Cano, López-Meneses and Sarasola, 2013) (Figure 4.7).

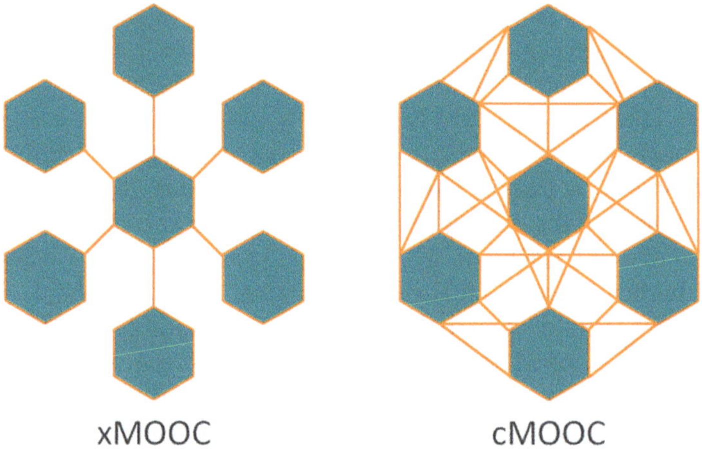

Figure 4.7 xMOOC vs cMOOC.

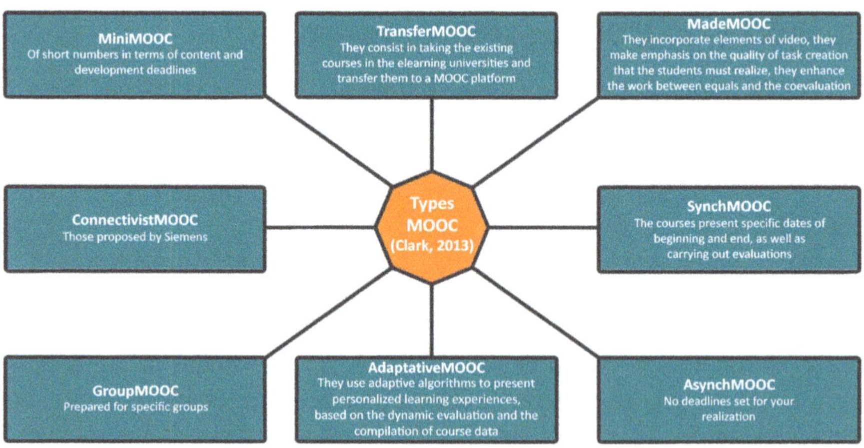

Figure 4.8 Type of MOOC courses.
Source: Clark (2013). Own elaboration

However, it is also known the different categories of MOOC enunciated by Clark (2013), can be seen in Figure 4.8.

XMOOCs tend to be traditional university e-learning courses that adapt to the characteristics of MOOC platforms, while cMOOCs are guided by the connectivist learning guidelines of George Siemens and Stephen Downes (Figure 4.9).

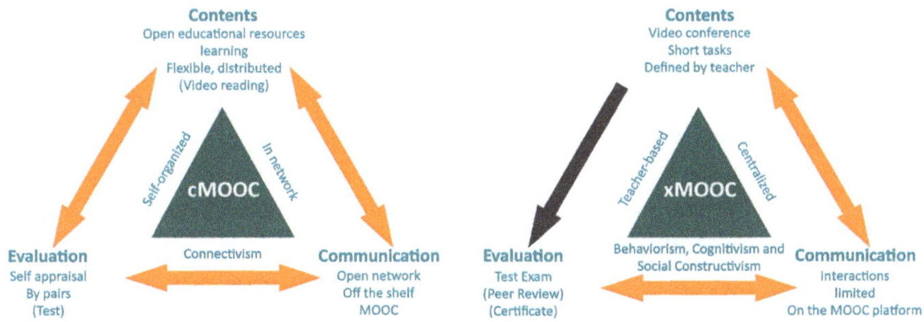

Figure 4.9 Central ideas of cMOOC and xMOOC. Adapted from Yousef et al. (2014).

4.2 Characteristics of xMOOC Courses

The xMOOCs are models of courses that adopt a more behavioral approach, emphasizing individual learning versus peer learning (Conole, 2013) and in which students acquire a series of content. To a certain extent, it could be said that they are the same online versions of traditional learning formats (reading, instruction, discussion, etc.) that universities develop in their e-learning actions. In them, the development of the course is more similar to a traditional course where the interaction with the rest of people is in the background and where the student receives a significant amount of structured and sequential information, and is subsequently evaluated. Likewise, xMOOCs are more focused on the incorporation of new educational methods and technologies in their platforms, in addition to offering more and better contents in the courses.

They usually start from a more closed virtual space, a web page as a portal where you have to register and prove your identity and although it is the teaching teams that prepare the contents and plan a learning itinerary, there is a certain institutional planning in place. conception and design, guaranteeing minimum quality since the institution itself is benefited or criticized by opinions regarding the quality of the courses offered (Martínez de Rituerto, 2014). Also, normally in its design, video-plays take on a very significant role as an element of presentation of the contents, and more specifically, the video-classes consisting of exhibitions of teachers supported by presentations in "*PowerPoint*" or "*Prezi*" (Vázquez-Cano, López- Meneses and Barroso, 2015).

From the theoretical point of view, the instructional model is very close to the behavioral model (E-R), in which the stimuli are represented by the contents presented in the videos and other materials, and the responses are expressed by the students in the evaluation tests, whose immediate results act as positive reinforcers of the individual's learning behavior; a similar effect has badges or online badges that are awarded to students for participating in certain activities of the course (Figure 4.10). This model is used by some well-known MOOC course aggregators such as Coursera, EdX, Udacity, Khan Academy, and MiriadaX (Ruiz-Bolivar, 2015).

The xMOOCs are the type of MOOCs that are recently being discussed at an educational level and are probably those with the highest number of people enrolled (Martí, 2012). Also, they postulate themselves as a true type of business model, either by favoring the creation of a "*brand image*" of the university, being easy to control, or by certifying the course to users. The training actions of EdX, Coursera, and Udacity are based on this type of design.

For Yousef et al. (2015), the xMOOCs are characterized by the replication of traditional education practices used by formal educational institutions and focus on the provision of high quality content and follow instructional design methodologies directed by teachers. In addition, xMOOCs provide flexible access to a wide variety of teaching materials and offer the opportunity to

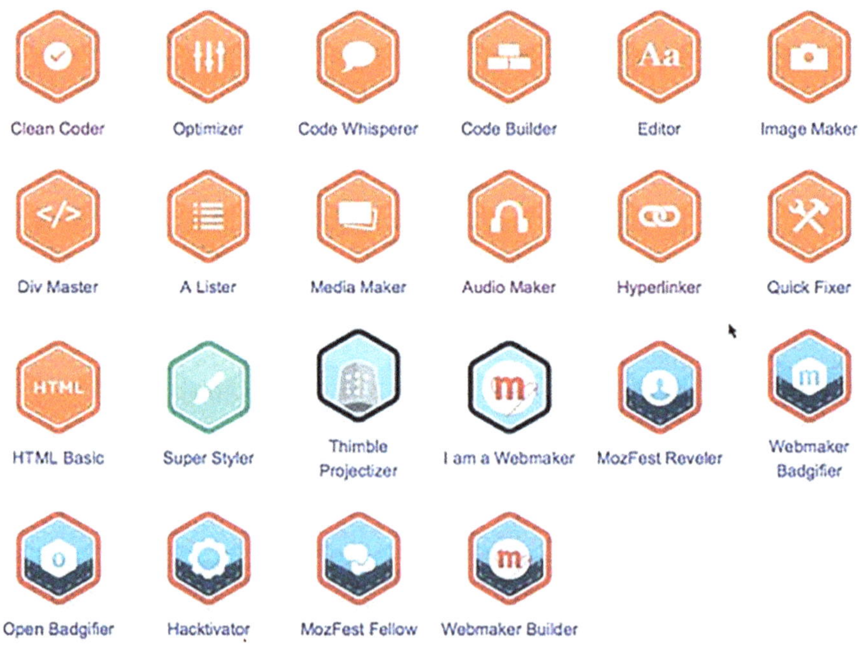

Figure 4.10 Examples of badges.
Source: Mozilla Foundation's Openbadge Project. URL: https://openbadges.org

combine face-to-face teaching with online teaching. Likewise, they seek to ensure that students acquire a series of contents and tend to be the same versions of the courses in e-learning but located in the specific platforms of the MOOC (Cabero, 2015). About them Vázquez-Cano, López-Meneses and Sarasola (2013:33) point out that "*the great problem of this type of MOOC is the treatment of the student in a massive way (without any type of individualization) and the methodological format already overcome of the trial -error in the evaluation tests.*"

In this proposal of the xMOOC, the function of the faculty is that of an expert, which selects the contents that must be transmitted to the students, and that of constructing the items that will make up the standardized and automated evaluation tools that this will have to overcome in order to acquire the certification to pass the course. As can be seen, the evaluation model used is very similar to the one followed in the traditional virtual training classes, where from a wide database of questions, different exams are randomly constructed for the participants.

In summary, as Valverde (2014) points out in the didactic proposals of the xMOOC, the fundamental axis on which all design and curricular development revolves is knowledge as a product (pre-packaged content), unidirectionality in the transmission of the contents that it gives to the role of expert person that the teacher adopts, and the "banking" conception of education in which "knowledge is a donation of those who judge themselves.

4.3 Characteristics of cMOOC Courses

One of the theories of learning that is related to this type of cMOOC course is connectivism (Siemens, 2005, 2008, 2010), hence the "c" (from "connectivism") with which reference is made to it. From this theory of learning, knowledge is not focused on the faculty, on the experts, but on the interactions and connections that the students who participate in the formative action establish among themselves (Moya, 2013), and it is precisely from these connections that it is shown that the participants establish how the learning is achieved. It is based, from the theoretical point of view, on the connectivism proposed by Siemens (2005), whose principles are:

- Learning and knowledge depend on the diversity of opinions;
- Learning is a process of connecting nodes or specialized information sources;
- Learning can reside in non-human devices;

- The ability to know more is more critical than what is known at a given moment;
- The feeding and the maintenance of the connections are necessary to facilitate continuous learning;
- The ability to see connections between areas, ideas and concepts is a key skill;
- The update (precise and current knowledge) is the intention of all connectivist learning activities;
- Decision making is, in itself, part of the learning process;
- The act of choosing what to learn and the meaning of the information that is received is seen through the lens of a changing reality. A correct decision today may be wrong tomorrow due to alterations in the information environment that affects the decision.

According to Wade (2012), he considers that connectivism can be seen as a theory that provides guidelines for instructional development within an educational context due to the notion that learning lies in building and connecting the knowledge that is distributed in a network of connections.

Unlike the previous model, the contents presented in them are minimal, and one of the functions that students must play in them is the search, location, and mix of information. Information that can depart not only from the documents of the environment, but also from the PLE themselves[1] of the students (Ruiz-Palmero et al., 2013; Cabero, 2014).

In turn, cMOOCs do not focus so much on the presentation of content in a formalized way, but rather on discursive communities that create knowledge jointly (Lugton, 2012). The MOOCs designed under this perspective are based on distributed learning in the network and the connectivist theory and its learning model (Siemens, 2007, Ravenscroft, 2011).

In this type of courses, different personal learning environments are used, scattered over the Internet. The necessary learning ends up being obtained from a global puzzle of knowledge pills disseminated anywhere in the Network that has an interest for the students and that have been located, analyzed, criticized, and democratically supported as valid by the people who participate in the course. In addition, these proposals are analyzed critically by each member of the group, and in some cases are rejected or modified and expanded by the detection and staging of new open resources agreed by the group itself, increasing and thus improving the learning proposal initial (Martínez de Rituerto, 2014).

---

[1] Acronym of Personal Learning Environments.

In this way, it focuses on gaining meaning from the learning experience with others. Therefore, while in the xMOOC, the control is located in the design process of the training action, and in the presented contents, in the cMOOCs, the students acquire a fully significant role in their own training process, and in them the interaction becomes a key element for learning to be achieved. Thus, students are increasingly self-taught as they learn and unlearn in the promotion of interaction and collaboration with other people in the group (Peláez and Posada, 2013; Pérez Parras and Gómez Galán, 2016b).

The significance that people have in this model of action and design of MOOCs leads to indicate if in xMOOC what is important is the institutions, in the cMOOCs what is really significant are the people, the contributions they make regarding the topic that is being analyzed, and the collaborative discussion they carry out with the rest of the participants for the construction of knowledge (Martí, 2012). They are therefore based on the idea that learning is generated, thanks to the exchange of information and participation in joint teaching and through the intense interaction facilitated by technology. In this sense, one would be speaking therefore of a formative action with a certain similarity to a social network of learning (Cabero and Marín, 2014). In short, in line with Peláez and Posada (2013:180): "they focus on gaining meaning from the learning experience with others".

On the other hand, unlike the xMOOCs, the cMOOC is a new learning space, in some cases annoying and uncomfortable, which is entering the universities in a disruptive way, which clashes not only with the way of teaching but also with the canons and the business model of the university itself (Vizoso-Martín, 2013).

Once its structure is seen, it is perceived that it is a model where traditional evaluation becomes very difficult and, therefore, it is usually assumed based on evidence. In this sense, it is assumed that learning fundamentally focuses on the acquisition of skills by the conversations that are generated, underlining an evaluation through evidence (Cabero and Marín, 2014). Following the same authors, the design of these courses follows clearly different approaches to the previous ones, since the structure is supported by the presence of external facilitators, use of a wide range of materials, and the promotion of high levels of apprentice control about the modes and places of interaction. In general, there are usually weekly synchronic sessions with facilitators and guest speakers, exchanges of information through forums, blogs, and social networks, and the use of a variety of resources such as concept maps, videos, multimedia presentations, and podcasts.

This structure has led some authors to consider the cMOOCs are replete with very chaotic training activities, since they are not prescriptive, and the people who participate in them establish their own learning goals, and the type of commitment they will follow, which ultimately has repercussions in that these students do not necessarily acquire a fixed knowledge and/or preconfigured skills in their beginnings (Lugton, 2012). Therefore, these MOOC courses are extremely complex when establishing fixed competences for all students and do not integrate well in the conception of obtaining a certification after completion where the competences are specified, since each person determines his/her competency and process and progress depending on his/her connections. The evaluation model used is not as formalized and regulated as it is with xMOOCs, and herein, the evaluation by pairs, the evaluation by evidence, and other more open models of evaluation acquire greater relevance (Cabero and Marín, 2014).

In general, for a correct performance in the MOOC courses, the participants must possess a series of competences, such as high technological and instrumental competences, high digital competence, a strong level of autonomy for learning, and high competence in self-regulation of learning (Vázquez-Cano and Sevillano, 2011, Cabero and Marín, 2012). In the cMOOCs, in addition to these, the desire to work jointly and collaboratively and in social networks must be incorporated, an aspect that as different investigations are showing is not always well perceived and admitted by the students (Cabero and Marín, 2013).

These two major types of MOOCs enhance the acquisition of different competences by the participants and allow different types of competences and capacities to be perfected. Moya (2013) has tried to establish differences between both types of designs from the four pillars of education pointed out in the Delors Report[2] (1996, p. 34):

- learn to know;
- learn to do;
- learn to live together and learn to live with others; and
- learn to be.

In Table 4.1, the author's vision of the significance that these different types of MOOC designs offer in this regard is offered.

[2]Report to UNESCO of the International Commission on Education for the 21st Century, chaired by Jacques Delors: http://www.unesco.org/education/pdf/DELORS_S.PDF

Table 4.1 Comparison between xMOOCs and cMOOCs based on the pillars of the Delors Report Education

Pillars of Education	xMOOC	cMOOC
Learning to know	• The learning centered on the information transmitted by the teacher. • Linear and guided learning	• Learning from sharing knowledge with others. • Active and participatory learning.
To learn to do	• The tasks they propose are more to assess if they have assumed the contents from a self-assessment. Learning is passive.	• The tasks depend on the involvement of the participants and their relationship with the rest. • It is a more active learning, highlighting learning by doing: "*learning by doing*".
Learning to live together	• From the approach of the xMOOC model, this perspective of learning to live together is not contemplated, since the learning process is totally individual.	• The connection established in this type of course is a good example of shared, collaborative, cooperative learning and, therefore, implies a relationship with the rest of the course community.
Learn to be	• The xMOOCs propose a totally individualized learning; hence, they will depend on the participant who develops or not. • Character of formation and learning for all the life: "*long life learning*".	• The proposal clearly reflects this learning, since it implies at all times that the connection with the rest of the participants and the interactions make us grow and develop as people. • Maintains the essence of lifelong learning: "*long life learning*".

Source: Moya (2013).

The fundamental distinction between both models of MOOC can be seen in Table 4.2.

For our part, in agreement with Bartolomé and Steffens (2015), we share that cMOOCs have a greater potential than xMOOCs to promote learning and self-regulation since they foresee a greater degree of interactivity with learning objects, peers, and tutors. In addition, cMOOCs are virtual learning

Table 4.2 xMOOC vs cMOOC

xMOOC	cMOOC
Traditional approach	Network learning
Linear	Non-linear, chaotic
Conceptual learning	Individual learning
Common contents	Distributed knowledge
Not scalable	Scalable network
Passive learning	Active learning

Source: Vázquez-Cano; López-Meneses and Barroso, 2015.

environments in which participants are active in acquiring, sharing, and creating knowledge, while xMOOCs focus only on providing knowledge.

Ultimately, the movement has opted, for the time being, for its materialization as xMOOC; what represents a more encapsulated training model than a commitment to participation, collaboration, and competency learning. Likewise, the MOOC movement has to overcome a series of difficulties that ensure its future sustainability, among which stand out: the pedagogical design, economic management, or "monetization", the certification of the studies offered, the monitoring of the training, the authentication of the students, the "Americanization of the movement" and the competence approach of its development.

Not to take the risk, to convert this type of training in another business type, such as "McDonald's", imbued by an Americanization of training and culture. The movement must overcome a pedagogical model encapsulated in an "impoverished e-learning" and move towards more collaborative and competent models, taking into account the cultural and linguistic diversity of different socio-cultural areas and contexts (Vázquez-Cano and López-Meneses, 2014).

5

MOOCs: New Learning Alternatives for the Expansion of Knowledge

At present, we find a technology that is behaving like a real earthquake, a dinosaur that destroys everything and that transforms everything, the MOOC. This is leading many people and institutions to consider them as part of a technology that in the near future will transform education (Vázquez-Cano, López-Meneses and Barroso, 2015).

Traditionally, university education has been based on a methodological model centered on the teacher, with emphasis on the transmission of content and its reproduction by students, the master class, and individual work. Teaching through Information and Communication Technologies (ICT) demands a series of changes that generate a rupture in this model (Ramírez, 2014a, 2015a, 2015b), at the same time as they represent an advance towards the quality of Education University (Aguaded, López-Meneses and Alonso, 2010a).

It is often heard that today's youth is not the way it was before. If there is a feature that characterizes the youth today, it is the fact that they were born in a technological world (Chiecher and Lorenzati, 2017). In this sense, talking about the processes of teaching and learning at the present time implies not forgetting, taking up the idea of Bauman (2006), that we live in "liquid times", where everything is dynamic and changing, and nothing is stable and solid. And these liquid times have an impact without a doubt on the transformation of the educational phenomenon in a number of aspects. Aspects that range from the emergence of new theories of learning (invisible, rhizomatic, ubiquitous, etc.), the speed and immediacy with which information appears and disappears, finding ourselves in a network society, moving in a society where the lifelong learning process is absolutely necessary, the empowerment and disappearance of certain cognitive skills, the breadth of information and communication technologies with which we find ourselves to carry out this process, and the need to have a high competence in multiple literacies (Cabero, 2015).

The evolution of distance education and technological advances constitute an important opportunity to increase access to education and contribute to the fulfillment of international educational commitments. In recent years, they have been characterized by a formative commitment based on activities, courses, and proposals based on technology-mediated teaching-learning processes (Castaño-Muñoz, Duart and Teresa, 2015; Estévez and García, 2015; Roig-Vila, Mondéjar and Lorenzo-Lledó, 2016; Colorado, Marín-Díaz and Zavala, 2016; Viberg, and Grönlund, 2017; Söllner, Bitzer, Janson, and Leimeister, 2018).

In this techno-social field, the emergence of MOOC courses in different areas of Education and Training has been unstoppable (Cormier and Siemens, 2010, Siemens, 2012, Downes, 2012a, 2013, Yuan and Powell, 2013, Dillenbourg et al., 2014; Daniel, Vázquez-Cano and Gisbert, 2015). In this sense, at the end of the first decade of the 21st century, with the arrival of MOOCs, a new educational panorama is emerging, posing new challenges to teaching and learning due, fundamentally, to its characteristics of mass, ubiquity, and gratuity. In turn, this phenomenon, which emerged in the United States, is attracting great interest among academics and policy makers in Europe (Sancho-Vinuesa et al., 2015). For that reason, Martínez-Abad et al. (2014) analyzed the impact of the word "MOOC" against "e-Learning", based on an analysis of scientific databases. The study shows that MOOCs are booming at the scientific level, with a significant increase in the number of publications, but that until now these have a more informative than academic slant, due to the relative short time it takes the phenomenon in action.

The UNESCO (2013) considers that these massive open and free processes are an opportunity to provide training in different contexts and the possibility of guaranteeing life-long training to the world's population. Likewise, in open learning, teachers will generate learning opportunities of individual and collective character in contexts of access and generation of content that develop tasks and skills and promote in the student body skills and competencies in accordance with the standards of higher education (Dublin Descriptors, 2005, European Commission, 2007, European Commission, 2008, Villa and Poblete, 2007).

On the other hand, due to its philosophy, these courses extend worldwide, being a clear example of disruption (Anderson and McGreal, 2012; Conole, 2013; Vázquez-Cano, López and Sarasola, 2013; Cobos Gómez Galán, and Sarasola, 2016) due to their cost, the number of students that it admits and its adaptation to the new social needs with respect to education. In turn, these have been understood as the latest evolution of online learning (Castaño et al., 2015) and have captured this worldwide interest due to its great potential to offer free training, offered by prestigious universities and accessible to anyone, regardless of their country of origin, their previous training and without the need to pay for their enrollment (Daniel, 2013, Christensen et al., 2013, Radford et al., 2014, Aguaded, Vázquez-Cano and López-Meneses, 2016).

MOOCs as new *massive didactic paths* can be constituted in a new techno-social tendency, especially oriented in the panorama of higher education to stimulate university innovation, or, simply, derive towards a new business model for Universities and Institutions without a quality demonstrated (Zapata, 2013; Vázquez-Cano, López and Sarasola, 2013). Without forgetting, in line with Martín, González and García (2013), that there is still a deficit of research on the evaluation of this movement.

Ultimately, this new training modality is generating more questions than answers in the scientific community (Raposo-Rivas, Martínez-Figueira and Sarmiento-Campos, 2015). Will they be viable in the future? Why is there so much interest in them? Will they transform the future of e-learning? (Gómez Galán and Pérez Parras, 2014 and 2017; Sánchez, León and Davis, 2015; Ramírez-Fernández, Salmerón and López-Meneses, 2016). In the following lines, the discrepancies and questions about the pedagogical value and the scope of this renovating and creative movement in the socio-educational context will be described.

Perhaps because in the first instance, the *Massive Open Online Courses* in this panorama of open and free education arise as a need for specialization that does not entail accreditation or certification as a priority objective, but that favors an approach to new labor and scientific realities, that the proposals of regulated education more corseted cannot offer (Vázquez-Cano and López-Meneses, 2015). In this way, they are offered by prestigious universities, opening new pedagogical paths for the massive expansion of global knowledge (Vázquez-Cano, López-Meneses and Sarasola, 2013).

Nowadays, the impact of mass, online, and open courses has been unquestionable in the recent years (Cormier and Siemens, 2010, Siemens, 2013, Downes, 2012a, 2013, Yuan and Powell, 2013, Dillenbourg et al., 2014). MOOCs have strongly broken into the context of higher education, and the movement is already considered one of the biggest disruptive events in the recent years (Pappano, 2012, Anderson, 2013, Conole, 2013, Conole, 2015, Little, 2013, Aguaded, 2013, Duart, Roig-Vila, Mengual, and Maseda, 2018).

In this sense, as indicated by Aguaded, Vázquez-Cano and López-Meneses (2016), without a doubt, the MOOC movement has been a turning point for higher education. But, for its definitive consolidation, a strong research component is needed that analyzes from different positions the principles on which the movement is based and, especially, under those aspects that present the most controversy: the pedagogical model in which the technological challenge is sustained that implies giving an adequate response to the massiveness and the challenges before certification and monetization that guarantee future sustainability and its definitive settlement in the formative panorama of higher education (Eaton, 2012; Zapata, 2013; Aguaded, Vázquez-Cano and Sevillano, 2013, Guàrdia, Maina and Sangrà, 2013, Sangrà, 2013, Hoxby, 2014, DeBoer et al., 2014, León et al., 2017).

Mackness, Mak, and Williams (2010) emphasize that MOOCs are modern means of teaching and learning, with a high potential for the exponential propagation of knowledge, because they are based on social networks or virtual learning environments. They also expand access to training by offering learning opportunities regardless of affiliation to a particular institution (Durall et al., 2012). Along the same discursive line, Marauri (2014, p. 40–41) justifies the development of MOOCs in the teaching processes because:

- They are very interesting because they disseminate knowledge among society, reaching new audiences and improving the reputation of institutions, which in this way are advertised as innovative entities and sources of high quality knowledge;
- Public institutions return to society in this way the investment that society has made in them;
- They allow anyone to continue training throughout their lives in a very specialized way and have new learning experiences for free, whether their ultimate interest is to obtain an accreditation or recognition, or only to be trained conveniently before a need or intellectual concern;
- It serves the teaching staff as a way to promote their teaching activity and their publications and in this way attract new students to regulated courses and permanent and continuous training. They also manage to increase their invitations to conferences and congresses by being the most popular authors;
- As they are free and open, no prior academic requirement is needed. Although it is always left in the hands of the different institutions and teaching teams to indicate and set the levels or minimum previous requirements necessary to be able to take them with guarantee of success.

From a positive point of view, these massive courses, through a process of systematic development could generate teacher training processes, both initial and ongoing. The massive nature of this type of training can mark a before and after in the coverage of the needs of teachers, especially in Africa and Asia, where it is most needed (Silvia-Peña, 2014).

With regard to design recently, Guo, Kim, and Rubin (2014) point out some characteristics that videos must have in these MOOC courses: these must not exceed 6 minutes in order to be more effective; the videos that intersperse the image of the teacher in a multimedia presentation are more effective than those that show only the presentation; those where teachers are drawing on a blackboard are usually more effective than those that contain a simple presentation; the master classes recorded in videos are not very effective even if they are divided into different short parts; and the behavior of the teachers in the video developing the session in an agile way and showing their enthusiasm for the subject are more effective.

Within the context of higher education, MOOCs could be new avenues for the expansion of knowledge, university innovation, employability, and the sustainable development of massive learning scenarios for global citizenship (McAuley et al., 2010; Méndez García, 2013).

Although there is still a long way to go (high abandonment rate, reflection on the most appropriate pedagogical model, organizational and business models, dominant culture, optimal instructional design, educational gap, among others), MOOCs contribute to social inclusion, the dissemination of knowledge, and pedagogical innovation, as well as the internalization of higher education institutions (Teixeira et al., 2016). It is also a place of common knowledge creation (Mañero-Contrera, 2016) and as De la Torre (2013) points out they lead one to recognize the significance that informal learning has in modern society, since it is increasingly important in the labor market, and demonstrate the "ability to do new things" instead of "the things you are able to accredit", and opens a path to learning and a path of connection and collaboration (Vizoso-Martín, 2013).

Ultimately, this new form of knowledge expansion in open, mass and online, which is part of the new educational fabrics of most international prestigious universities, can be erected as a transforming element of the classrooms, limited in time, spatially limited and sometimes reserved for a social elite; and thus transcend new ubiquitous, connective, informal, and horizontal learning scenarios that can facilitate the digital inclusion of the most disadvantaged people and the birth of interactive virtual habitats of free learning and collective intelligence and the sustainable development of homo-conexus formation/digitalis on the path of globalized knowledge (Ramírez, Salmerón, 2015, Vázquez-Cano, López-Meneses, E. and Barroso, 2015; Backhaus, K., and Paulsen, 2018).

In the face of initial illusionism, data is beginning to appear that calls attention to the fact that MOOCs, neither as an educational model nor as a business model, have a clear positioning, and that there are many failures in their implementation and abandonment by the people who participate in them. These facts should make us reflect on some of the decisions adopted (Chamberlin and Parish, 2011, Scopeo, 2013, López Meneses, Vázquez Cano and Gómez Galán, 2014).

The challenges, which could be called "traditional", of online education: the design of activities, facilitation, evaluation, and feedback (Burkle, 2004, Prendes, 2007, Sánchez-Vera, 2010) maintain and even intensify with the MOOC, since the massification of the courses makes these tasks even more difficult (Sánchez, León and Davis, 2015). And, in some MOOC courses, the transmission of the content by the teacher or expert person predominates, promoting the traditional teaching model (Zapata, 2013, Hollands and Tirthali, 2014, Cabero, 2015, Valverde, 2015).

Another weakness of this type of online and massive training is the strong dropout rate of the MOOC courses (Zapata, 2013). Likewise, it is corroborated in different investigations, such as the study of Gee (2012), reflecting that the rate of completion and improvement did not reach 5% in the development of a course of the Massachusetts Institute of Technology (MIT). Similarly, the Armstrong study (2014) states that only 4% of the students enrolled in a MOOC of the Coursera platform completed their courses. On the other hand, Cabero (2015) points out that this high rate of failure, or more correctly massive abandonment, can be due to a series of variables, one of them possibly being the different typology of people who decide to start this type of training activities.

In this sense, Hill (2012) indicates the possible profiles of students enrolled in a MOOC course:

- Non-participants: they register but after they do nothing, they only register;
- Observers: students that make up the bulk of participants in a MOOC and who, for the most part, only review a few elements of the course. Moreover, a significant number of these people hardly perform more than registering;
- Marauders: students interested in certain specific parts of a course, which will be the ones that review, leaving aside the rest of the material;

- Passive participants: students who only watch the videos and do the odd test, without getting involved in all the possible activities that the course offers (blogs, forums, p2p ...);
- Active participants: students totally committed to the course, participating in each and every one of the activities proposed by the teaching team, trying to make the most of the experiences that this new type of learning brings.

According to Cabero (2015), a possible solution could be to carry out initial diagnostic tests on the students, in order to know if they possess the initial competences necessary to be able to approach the course with a guarantee of success, or to indicate that for its realization, it is necessary to carry out some actions in advance. This would lead us to change the option of mass courses for "personalized open courses" (POOC). Likewise, following this same author, the pedagogical romanticism that has prevailed over these resources has made us forget that students must possess for the follow-up of the courses a series of competences, among which we could point out: high technological and instrumental competences (Cabero, Marín and Llorente, 2013). In other words, it is essential to have acceptable levels of digital competence, autonomy, and self-regulation of learning (Cabero, 2013).

Also, this high dropout rate in this type of course could be due, as indicated by Vázquez-Cano, López-Meneses, and Sarasola (2013), for its design and behavioral court structure, where all the sense of the formative action lies in the presentation of the contents. Without forgetting that part of the student who participates in the MOOCs is shown with some disorientation and overload have low probability of interaction with experts or guidance, little real socialization, and difficulties to follow the discussions that are maintained in social networks: Facebook, Twitter, Google groups, etc. (Chamberlin and Parish, 2011, Calderón, Ezeiza and Jimeno, 2013; López-Meneses, 2017).

One of the biggest challenges of MOOCs, as Vázquez-Cano (2015) indicates, focuses on the possibilities offered by technology to address mass education. It is a utopia to think that a course that can be done by more than 100,000 people can be attended and developed by two or three teachers, or through the traditional resources of online teaching. The integral treatment of the pedagogical model and the model of tutoring and evaluation of MOOC courses requires technological tools that enable an adequate treatment and according to the massiveness that occurs in these training environments. Also, as already noted, some authors have indicated that this may involve a "McDonalization" of education, generating a strong uniformity and an accentuated globalization of training actions, thus breaking any process of educational contextualization, thus facilitating a new process of cultural colonization (Hayes, Wynyard, and Mandal, 2002; Lane and Kinser, 2012; Popenici, 2014; Hayes, 2017).

Another debate focuses on the critique of the didactic and pedagogical model that underlies a course where technology is limited to viewing videos and answering a series of questions (Sloep, 2012; Rees, 2013):

- With this system do you really learn?
- Is it different from other training models?
- Can you learn and exchange opinions when more than 20,000 students can be enrolled?

In this sense, for this model to be consolidated, a pedagogical structure similar to the five phases indicated by Waite et al. (2013): introduction, reflexive practice, methodological practices aimed at group learning, information recovery, restructuring, and creation and evaluation (Figure 5.1).

Vázquez-Cano, López-Meneses, and Barroso (2015) express that the pedagogical and didactic models must be reoriented depending on the themes, the competences involved in the training action and the average number of people who enroll. Otherwise it could become a business for some companies,

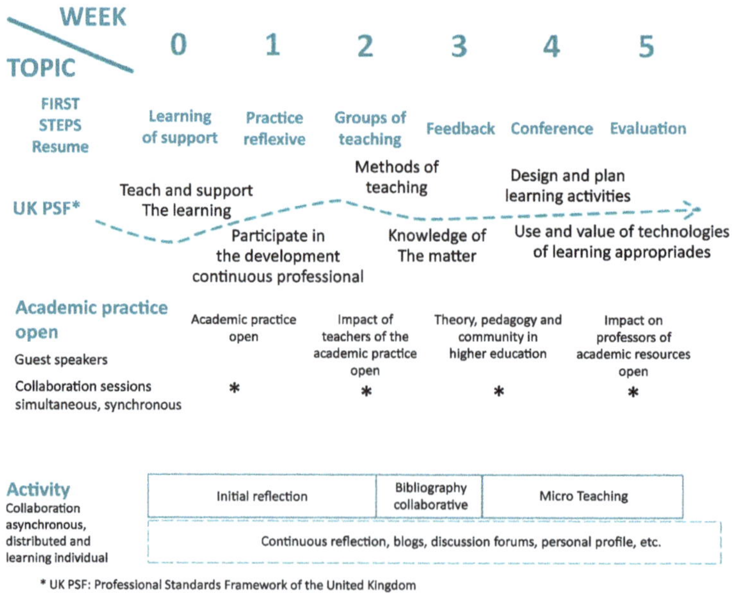

Figure 5.1 Pedagogical structure of MOOCs.
Source: Waite et al. (2013)

governments, and institutions at the expense of impoverishing the educational landscape (Mazur, 2012, Daniel, 2012).

Also, voices have been presented that speak of the aforementioned process of "MacDonalization" of education through the distribution of standardized educational packages worldwide (Lane and Kinser, 2013; Aguaded, Vázquez-Cano and Sevillano-García, 2013; Hayes, 2017). Obtaining a degree from these new technologies from anywhere in the world can be the friendly face of a process of transnationalization of universities. We are not only talking about the export of knowledge that could be leveled to the processes of ideological and cultural domination, but now we are talking about the process of obtaining a degree or a degree in the different institutions that have their headquarters in Anglo-Saxon countries.

Similarly, standardization can imply an excessive unidirectionality of knowledge (Lane and Kinser, 2012). In this sense, Chamberlin and Parish (2011) state that due to the large number of users enrolled in these courses, it is difficult to carry out meaningful interactions for learning, with added concern, as pointed out by Jason Lane and Kevin Kinser (2013), of being able to become a process of Americanization of the formation, since in a principle the elaboration of these had the intention to spread the institutions and a model of American formation throughout the world.

The evaluation of learning is a central feature in the pedagogical design of MOOCs (Sandeen, 2013). It is estimated that electronic evaluation in open courses has special characteristics, added to those of certification and accreditation of MOOCs due to their open nature. It would be necessary to avoid certain errors that have been committed in the design of the evaluation of learning in the past. online courses (Shank, 2012), such as waiting for a bell-shaped learning curve, choosing an incorrect type of evaluation, insufficient evaluations or poorly written multiple choice tests. Also, as Cabero (2015) expresses as an evaluation, usually simple knowledge tests are used, not assessing the participation or interaction of the student with their peers and the teacher.

Other difficulties that must be faced in the design of these courses are the financing and subsequent commercial exploitation of the courses. At present, the university institutions are beginning to sign agreements with companies for the initial financing of platforms and open courses, but these investments are not carried out on a permanent basis but rather general strategies are established to monetize these initiatives (Vázquez-Cano, López-Meneses and Barroso, 2015). The two most used options are certification through badges and the sale of courses (Young, 2012). Also, the authentication of the identity of the participants to avoid deception, plagiarism, and personality impersonation is another problem that must guide the design of the MOOC courses (Wukman, 2012).

Regarding the certification of students who participate in these courses, as indicated by Vázquez-Cano, López-Meneses and Barroso (2015), a reconceptualization of the model could be made towards ways of certification and accreditation of the most innovative, flexible, and adapted to the needs of a labor market in constant evolution and growth, as far as professional profiles are concerned. In this sense, "badges", representing a skill or an achievement, as an iconographic and structured identification, should be designed based on the criteria that allow their emission and distributed circulation among related agents and collaborative work structures ("peer to peer"), which can be an interesting bet to continue advancing.

Finally, the supposed gratuitous nature of MOOCs can help overcome the difficulties of the economic, educational and productive system and the demand for skills adapted to the emerging market models. For this, these incipient systems of formation must overcome many deficiencies for a future sustainable construction, among which stand out: the pedagogical design, the economic management "monetization", the certification of the offered studies, the follow-up of the formation and the authentication of the students (Eaton, 2012; Hill, 2012; Touve, 2012; Egloffstein, 2018).

In relation to the design of activities offered by MOOCs, these must be oriented towards reflecting on the practice itself and the acquisition of new competences rather than the content instruction and the evaluation of them. Many of these courses do not go beyond offering training based on a traditional class, segmented into audiovisual presentations of no more than 15 minutes and in which the competence level of the student body is diminished by relying almost exclusively on rote-conceptual learning and on a mechanical evaluation of "trial-error" (Vázquez-Cano, López-Meneses and Barroso, 2015). In addition, there is the difficulty of the dispersion of information, conversations of the forums and interactions among hundreds of students that it is necessary to structure and organize for a holistic understanding of knowledge, that is, MOOCs need "*content curators*", experts who select, filter and systematize information continuously to help students enrich their learning process.

On the other hand, while it is justified as a great strength the fact that MOOCs facilitate access to knowledge to all social sectors (especially the most disadvantaged), this type of courses has not penetrated as deeply as could be expected in the populations of lower educational level and/or with difficulties to pay for a university degree (Christensen et al., 2013).

It is also necessary to highlight as weaknesses of these MOOCs the question of the language in which these courses are offered and the ability to be followed in non-English- or Spanish-speaking countries. It would be necessary to develop multilingual MOOC courses, closer to the emerging countries, together with MOOC courses more accessible to people with disabilities, for example, MOOCs adapted to sign language and technologically more compatible with mobile devices.

Another of current and future challenge of MOOCs includes articulating a feasible system of evaluation and certification of the competence progression of the participants in each course, such as the peer evaluation system, a more horizontal, networked and more related learning: the web 2.0. In this sense, different authors are investigating the need for quality criteria in the MOOC courses (Mengual, Roig and Lloret, 2015, Ramírez-Fernández, 2015, Aguaded and Medina, 2015; Stracke et al., 2018) and on student performance (Castaño, Maiz and Garay, 2015). The key, for a large part of the experts, is to achieve a system of data collection and annotation and of their analysis, combining hetero-evaluation with self-evaluation and peer evaluation (Cano, Fernández and Crescenzi, 2015).

According to Ramírez-Fernández (2015), MOOCs appear as the latest stage in the evolution of e-learning, and its quality is an emerging field for researchers and teachers in the university field with concern expressed by qualitatively and quantitatively measuring this type of training. In this way, the studies should focus on evaluating with tranquility what these courses offer in terms of their pedagogical value in the field of training through the Internet and, what is more important, how they can be improved in this regard (Aguaded, 2013; Guàrdia et al., 2013; Pérez Parras and Gómez Galán, 2015). In this same line, it does not seem so evident that MOOCs offer quality training (Martín et al., 2013) and it would be necessary to improve if they want to be a disruptive milestone (Roig et al., 2014).

Ultimately, as Bartolomé (2013) states, there is still a pedagogical reference framework that ensures that MOOC teaches and that MOOC learns. There will still be a need to refine concepts, models, experiences, etc. overcome difficulties encountered and minimize others; some MOOCs and platforms will stay on the way, but many others will continue to be designed, developed, and improved for millions of people in the world (Raposo-Rivas et al., 2015). In this sense, we agree with Aguaded, Vázquez-Cano, and López-Meneses (2016) that the MOOC movement, like any other science or field of study, requires recovery, evaluation, and analysis processes that provide the possibility of visualizing and represent in a comprehensive, consistent, relevant, and accurate way the result of their work and ensure the legitimacy and originality of the scientific knowledge produced.

In this way, the evaluation of the quality of the MOOC courses is on the research agenda for the future and the need for a greater number of studies on some evaluation quality indicators in online courses, as well as longitudinal studies (Stödberg, 2012) or comparative (Balfour, 2013), how can they be integrated into open learning environments (Oncu and Cakir, 2011) or focused on the most educational aspects (Roig, Mengual and Suárez, 2014; Raposo-Rivas et al., 2015, Gómez Galán and Pérez Parras, 2016a).

In short, among its weaknesses and limitations (Caballo, Caride, Gradaille and Pose, 2014, Cabero, Llorente and Vázquez, 2 014, Popenici, 2014, Valverde, 2014; Gómez Galán and Pérez Parras, 2017, Pilli, Admiraal, and Salli, 2018), highlight:

- Use of overcome methodologies: Focused on contents.
- It is massive. Lack of differentiated and personalized education.
- A certain pedagogical innovative romanticism.

A standardization of knowledge (MacDonalization of school culture).

- Lack of conceptualization and educational research.
- Devaluation of the teaching function in teaching and learning processes – Teacher evaluation as a unidirectional communicator. New paradoxical roles: monitors and facilitators.
- It requires a digital domain and self-regulated learning by the students.
- Large number of interactions that make evaluation and monitoring impossible.
- Rhythm marked by who designs.
- Inferiority by culture and languages.
- No tutorials and activities can become a repository of learning objects.
- They are led by fashion and the market.

And, according to Schulmeister (2012), they express the critical points of MOOCs in: lack of feedback and low interaction; there is no reliable verification of learning outcomes and peer evaluations and a wide variety of topics predominate, but without an explicit curriculum. And in high desertion rates (Fidalgo, Sein-Echaluce and García Peñalvo, 2013; Yamba-Yugsi, and Luján-Mora, 2017).

That said, but despite its difficulties, we believe that MOOCs open a new range of possibilities, since not only are resources released, but the entire educational process, and therefore represent another option to learn on the internet and expand our network of contacts, as well as represent very interesting training and professional update opportunities (Sánchez., León and Davis, 2015). They can even have advantages for teaching using them in Flipped Classroom experiences (Zhang, 2013).

MOOCs open doors to new learning opportunities throughout life (Kop et al., 2011, Bartolomé and Steffens, 2015), and it is a way to learn, ideally an open, participatory, distributed course and a learning network for all of life, it is a path of connection and collaboration (Vizoso-Martín, 2013). In addition, it is an incipient area of development that does not stop evolving and that is beginning to generate new areas of research (Vázquez-Cano, 2013).

As already mentioned, there is still a long way to go: high abandonment rates, reflection on the most appropriate pedagogical model, organizational and business models, the dominant culture, optimal instructional design, educational gap, among others (Vázquez- Cano, López-Meneses and Barroso, 2015). Overcoming these obstacles will enable the MOOC movement to become a new techno-social trend, especially oriented towards the higher education panorama for the stimulation of university innovation and the promotion of massive, open, and interactive learning scenarios for the genesis of collective research (Vázquez-Cano, López-Meneses and Sarasola, 2013).

Ultimately, the philosophy of this training modality implies a democratization of higher education (Finkle and Masters, 2014, Dillahunt, Wang and Teasley, 2015). In addition, they can mean, as we explained in Vázquez-Cano, López-Meneses, Méndez, Suárez, Martín-Padilla et al. (2013) the new educational paths irradiated in some of them with the spirit of educational innovation and education of quality for all throughout lifc. And despite its growing popularity and prominence, the most promising value of MOOCs does not derive from what they are, but from what they can become, that is, from the positive derivatives that are beginning to emerge and that derives from the flexible character and open learning that they advocate (Yuan and Powell, 2013).

In summary, in agreement with Natividad et al. (2015), the technology itself is neither good nor bad, and the great educational challenge is to make it effective, efficient and sustainable. And our desire is "to *participate in this rising socio-educational movement with many potentials and training possibilities but also with some challenges and difficulties that must be faced with reflection and scientific research, so as not to become a commodification of orphan titles of intellectual progress, but in true seeds for the global formation of Homo Digitalis*" (Vázquez-Cano, Méndez, Román and López-Meneses, 2013, p. 62).

6

Directory of Resources Related to MOOCs

MOOC platforms use the social media available to disseminate their activity and participate in social networks as done by the universities themselves (Cataldi and Cabero, 2010, Chamberlin and Lehmann, 2011, Túñez and García, 2012) to maintain an updated profile, promote the courses, the platform and interact with the users, and obtaining fast and direct feedback. This contributes to improving their corporate image (Kierkegaard, 2010), optimizing their service strategies and promoting their academic and professional activity.

Then, from a previous study carried out (Martín Padilla and Ramírez Fernández, 2016), a directory of the most used platforms is shown, taking as reference the Scopeo report (2013) that indicates that the main Anglo-Saxon platforms are Coursera, edX and Udacity, and the Ibero-American sphere the Miriadax platform. Also as indicated by different authors (Barnes, 2013, Jordan, 2014) are the most widely used by universities.

6.1 The Coursera Platform

Coursera is a platform of MOOC courses that was born in 2011, with the help of professors from Stanford University, with the aim of providing free courses around the world. Coursera is part of more than 148 educational institutions around the world, including some of the most prestigious universities, such as Princeton, Johns Hopkins, Brown or the Berklee College of Music, among many others (the complete list can be seen in https://es.coursera.org/about/partners). Among the Spanish universities are the Autonomous University of Barcelona, the University of Navarra, and the University of Barcelona.

The Coursera platform is available in different languages (English, Spanish, Portuguese, Chinese, French, and Russian). It has courses, both free and paid, in different languages, from English to Spanish, through French,

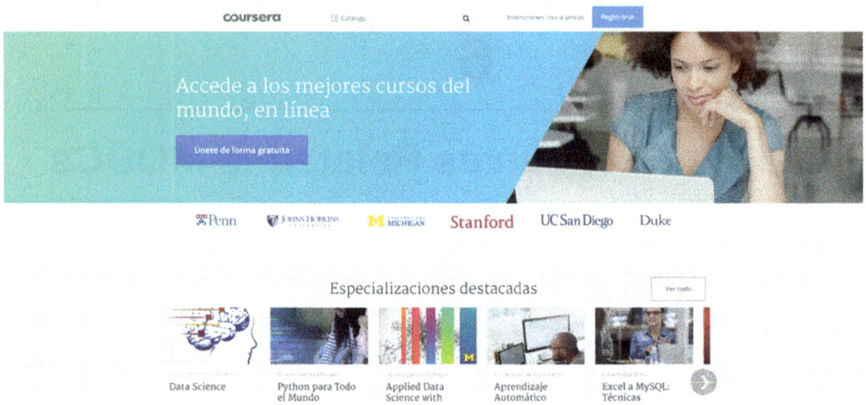

Figure 6.1 Coursera Platform home page.
Source: https://www.coursera.org/

Italian, Russian or Chinese, among others, which has courses in more languages than initially translated in his platform. It also offers the possibility of filtering the courses according to the available languages, after selecting the general category:

As for the themes offered, they are very varied and are included in the following general categories: Arts and Humanities, Business, Computer Science, Data Science, Biological Sciences, Mathematics and Logic, Personal Development, Physical Sciences and Engineering, and Social Sciences and Language learning.

Within these categories, we can find different areas and disciplines represented. Many of these training actions are introductory and others require some degree of prior knowledge in the study area to be able to follow their development with use. However, the level of deepening is not an information that is reflected in the catalogue, but, rather, is a matter inherent in the subject itself.

The registration on the page is free. Once the person has registered, they can access a list of available courses and perform the one they want. In some, they indicate the exact date in which they will begin, others are only planned in the absence of a definitive date. A few days before the start of the course, an email is sent to the registered people welcoming and explaining in a brief way the operation of the course. At the same time, access to the virtual course is enabled so that students can become familiar with the different parts of it.

The platform has several "Help Forums" in which you can request assistance with certain problems, whether they are of a technical nature or issues related to management. In addition, there are a series of "Help articles" that respond to the most common problems that users may encounter: problems with the account, account settings, problems with the use of certain web browsers, etc.

The structure of the course may vary, but the most common is to find the following design:

- *Course information:* This section provides information about the purpose of the course as well as the skills that students can develop. It also details the teaching team that has designed the course and mentors it, as well as the university or universities that support it. Finally, the course program and information on how to overcome it is included.
- *Preview of the course:* This section provides an overview of the planning of the course organized in weeks, indicating the number of multimedia resources that should be viewed each week and the questionnaires or tasks to be performed in each one. When selecting a specific week, access is given to the following contents:
    - Learning objectives: The objectives that are pursued with the content of the module are described.
    - Videos: Lessons recorded by the teaching team with explanations for each topic. Each video usually lasts between 10 and 15 minutes and each theme can contain several video lessons.
    - Forums: Forums for discussion and meeting between the students, teaching team and support team. Study groups are usually formed by languages, countries or interests. For example, groups of Spanish or "homeschoolers".
    - Questionnaires: Questionnaires are tasks that are automatically graded and used to assess student knowledge in a course. In some courses, the questionnaires are called "homework" or "exams" but both are useful for the students to demonstrate the acquisition of knowledge.
    - Tasks: We can find tasks of different types:
        - Testing tasks: These are tasks in which students must send a text following the established premises and that, later, will be reviewed by the teaching team;
        - Tasks reviewed by peers: This is a variation of the essay tasks in which the revision is carried out by the students. Each

student receives comments from other people who are doing the course, and each student should also review the tasks of other people and make comments;
- Honors assignments: Some courses have optional assignments that can be completed to obtain special recognition. They are not necessary to pass the course,and do not affect the eventual qualification.
- Programming tasks: These are tasks that require developing and executing a computer program to solve a problem and are automatically qualified. Some programming tasks influence the final grade of the course, while others are just practice;
- Mathematical tasks: These are questionnaires that allow you to write mathematical functions and certain constants using basic symbols.

Some courses require a high work rate for students. You have to visualize the videos, read the proposed texts, complete the questionnaires, perform the tasks, review and comment on the tasks of the rest of the students, etc.

The duration of the courses ranges from 4 to 11 weeks, and the ideal workload is 8 to 10 hours per week, although afterwards the students can organize the study time as they wish. An element that facilitates this aspect is that Coursera has decidedly bet on the possibilities offered by mobile learning and has an app for smartphones and tablets, for both iOS and Android. From the app, it is possible to make new registrations, or follow the courses in which we have registered, by "streaming" the video sessions of the course, or saving them for later viewing in "off-line" mode.

Regarding the help that can be received as a student, there is usually a good disposition on the part of the organization and the teaching teams to help and advise the students, especially in courses taught in languages that are not their native language.

The platform has the possibility of accessing through mobile devices using an app for both Android and iOS.

6.2 The edX Platform

edX is a well-known platform of MOOC courses created in 2012 jointly by MIT (Massachusetts Institute of Technology) and Harvard University. It is a non-profit and "open-source" project that, during these years, has joined more than 90 educational institutions around the world, such as the

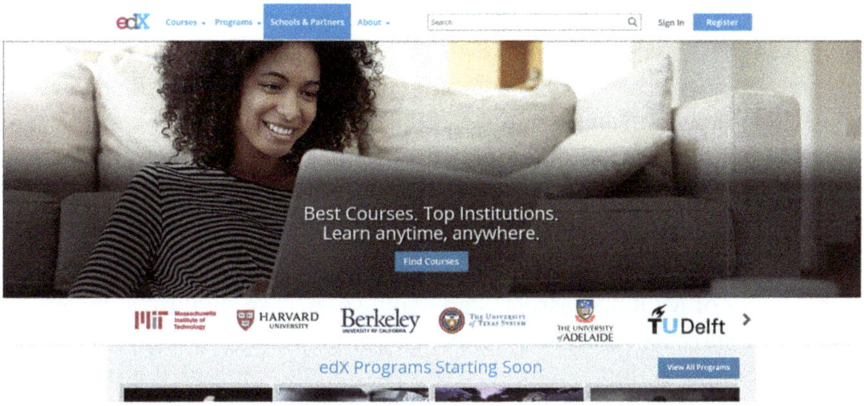

Figure 6.2 Start page of the edX Platform.
Source: https://www.edX.org/

University of Berkeley, the University of the Sorbonne in Paris, Princeton, the University of Oxford, etc. Among the Spanish institutions, the Polytechnic University of Valencia, the Carlos III University of Madrid, and the Autonomous University of Madrid, stand out. The full list can be accessed at: https://www.edX.org/schools-partners#members.

Being a non-profit initiative, access to the courses is completely free. However, if you want to receive some type of certification, you have to make payments, using the money collected for the maintenance and improvement of the resources that are made available to teachers and students. The courses offered are organized around the following subjects: Architecture, Art and culture, Biology and Life Sciences, Business and Management, Chemistry, Communications, Computer Science, Data Analysis and Statistics, Design, Economics and Finance, Education and Teacher Training, Electronics, Energy and Earth Sciences, Engineering, Environmental studies, Ethics, Nutrition and food, Health and safety, History, Humanities, Languages, Literature, Law, Mathematics, Medicine, Music, Philosophy and ethics, Physics, Sciences, and Social Sciences.

In its formative catalogue, edX offers very detailed information regarding the characteristics of its MOOC courses. From the data offered, it can be verified that most of them are in English (85.6%), with Spanish being the second language with the highest presence (7.7%), followed by Chinese (3.3%) and French (2.3%). There are courses in other languages (German, Japanese, Portuguese, etc.), but their number is almost anecdotal if we make a comparison with the previous languages. Among these courses we can find MOOCs with different levels of complexity. Most of the courses offered are for beginners (57.39%), although there are also intermediate level courses (32.24%) and advanced courses (20.27%).

The edX platform offers an excellent search engine for courses with different filter levels that serve different search variables, so it is very easy for students to find the course that best suits their needs. The variables by which a search can be filtered are the following:

- Availability of MOOCs: assets, start soon, soon, own rhythm, archive;
- Subject or subjects (according to the list commented on above);
- Type of training action: courses, programs, MOOC verified, with credits, micro-masters, etc.;
- By university or institution that organizes the MOOC;
- Level: advanced, intermediate, initiation;
- Language in which it is taught.

Regarding the resources available in each MOOC course, edX offers a very clear, simple and intuitive interface. In the upper part we find accesses to the different tools of the course:

- Start: When you access each course, you will be shown a homepage in the form of a summary in which news and information about course updates is provided. On the right side of the screen, information about important dates of the course and the content structure of the course appear as hyperlinks. Also shown on this screen is a link through which to contact the support team;
- Course: When accessing this section, the structure of the course is available in the left side menu. The courses are organized by week and within each week an organization is established by themes. To access the contents of each topic, a menu is available at the top of the screen, from which we can access the different videos, questionnaires, and tasks, as well as move to previous or subsequent topics. The usual thing is to alternate video and test questionnaire that refers to the contents worked on. The tasks can be questionnaire type multiple choice test or essay type tasks in which students must write a broader text;
- Discussion: Through this link, you can access the different discussion forums of the course. Usually, a discussion forum is established for each topic (or sub-topic) and a presentation forum or cafeteria type in which students can discuss other topics of their choice;

- Progress: This section shows the progress of the students regarding the development of the course. A general graph and detailed information is shown per week and, within it, by topics;
- Program: This section provides detailed information on the characteristics of the MOOC course: duration in weeks, objectives of the course, contents, prerequisites, information on the evaluation and weight of each task in the final grade, etc.;
- Frequently asked questions: On this page, answers to the most common questions about the operation of the platform and other questions related to the course are often collected in the form of explanatory videos.

The platform has the possibility of accessing through mobile devices using an app for both Android and iOS.

6.3 The Udacity Platform

Sebastian Thrun and Peter Norving, renowned professors of Artificial Intelligence at Stanford University, organized in 2011 an online course entitled "Introduction to Artificial Intelligence". The course was extremely successful, as it involved more than 160,000 people from more than 190 countries, a figure considerably higher than the slightly more than 500 people who attended the face-to-face classes they both taught at the university each year. This fact made Thrun fully aware of the possibilities offered by this new form of education, free and interactive. That encouraged him to leave his post at Stanford University and found, together with David Stavens and

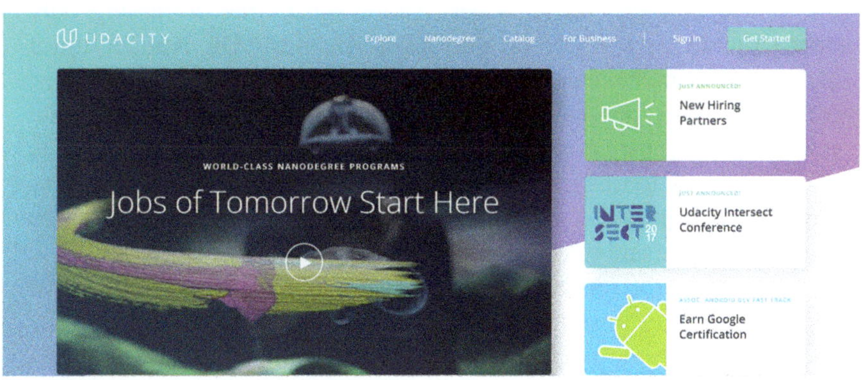

Figure 6.3 Homepage of the Udacity Platform.
Source: https://www.udacity.com/

Mike Rokolsky, a company called "Udacity" which they called the free online university. Udacity began its journey in early 2012 with an introductory course to search engines and throughout that year it progressively incorporated more courses related to different areas of knowledge.

As with other platforms, Udacity has associated institutions, although in this case, it is mostly companies and not universities, although there are specific agreements with some of them, such as the State University of San José (USA). Among the most prominent partners are multinationals such as Google, Amazon, Facebook, NVidia, IBM, or Mercedes-Benz.

The platform is available in English, Portuguese, Arabic, German, Chinese, Korean, Japanese, and Indonesian. In terms of the courses offered, they are mainly focused on topics related to technology: programming languages (Android, Java, etc.), Virtual Reality, Web Design, Artificial Intelligence, etc., although there is a category called "Non-Tech" (Not technology) where you can find some MOOCs related to mathematics, physics, psychology, biology, etc.

When selecting a course of interest, it is possible to carry out filters on different variables until a specific profile is established. The selection can be made in relation to the following variables:

- Category Thematic: Android, Data Science, iOS, Web Development, Non-Tech, etc.
- Type: free courses, postgraduate courses.
- Beginner, intermediate, and advanced levels.
- Company or entity that manages the course.
- Concrete technology that is worked on in the course.

The interface of each course is very short, since when accessing only the content structure appears in the side menu. Each MOOC is made up of different teaching units which in turn include various video lessons, to which subtitles can be added, question-type questionnaires in which questions are raised regarding the topics covered in the videos and which are intended to help the student body to understand concepts and reinforce ideas, as well as follow-up tasks that promote the "learning by doing" model, among which are those of essay type and those of portfolio.

Each MOOC course has a specific discussion space in which various topics can be addressed in relation to the contents worked on. It also offers a search engine within the interface itself and a section called "Resources", in which, among other issues, it is possible to download in compressed format (ZIP) all the videos included in the course, which may be of interest to see without Internet connection, or to dispose of them in the future. However, the access links to these tools do not appear sufficiently prominent, and may go unnoticed.

The platform has the possibility of accessing through mobile devices using an app for both Android and iOS.

6.4 The MiríadaX Platform

MiríadaX is the largest Spanish-Portuguese platform in the world, which saw the light in 2013. The project involves, on the one hand, Telefónica Learning Services, which is the Telefónica Group company specialized in offering integral online learning solutions for the education and training, and, on the other hand, Universia, the largest network of Spanish and Portuguese-speaking universities, promoted by Banco Santander. The project is supported by more than 90 prestigious Ibero-American universities and institutions and has more than three million students and more than 2,000 teachers.

The platform is presented only in two possible languages: Spanish and Portuguese. Regarding the topics that are worked on, we can find multiple topics: Anthropology, Astronomy and Astrophysics, Castilian, Political Science, Agricultural Sciences, Health Sciences, Earth Sciences and Space, Life Sciences, Sciences of the arts and letters, Legal Sciences and Law, Medical Sciences, technological, economic, Philosophy, Geography, History, Humanities, English, Linguistics, Mathematics, Pedagogy, Portuguese, Psychology, Sociology, and Ethics.

Figure 6.4 Home page of the MiriadaX Platform.
Source: http://miriadax.net/

The search engine of the course catalogue is less functional than on other platforms. It offers a search engine with which you can search based on a topic or keyword. On the other hand, the filtering is reduced to selecting a general topic or to carry out a filtering from the university or institution that sponsors the MOOC course.

The interface of the course is simple, clear, and intuitive. It has a support section where you can ask for help or access the Frequently Asked Questions and Answers of the students (FAQ). Also in this section, anyone can send their suggestions regarding the operation of the platform, or regarding topics that may be of interest.

When we access the MOOCs, we find in the upper right a menu with the different sections:

- Start: In this section, it usually shows an introductory video or presentation of the course and then the structure of contents. As usual in this type of platforms, each course is organized in modules, in order to facilitate its monitoring by students. These modules include publications, readings, and audiovisual material narrated by the teaching team in charge of the course in question. It also appears highlighted on the home page the university or institution that organizes the course, and the duration in weeks of said course;
- Syllabus: This section shows the structure of the course but focusing in a special way on the terms of completion of each module and the different activities and dates of the delivery of each;
- Notes: This section shows the results obtained in the activities that make up the course. It indicates date and time of completion in your case and approximate percentage of progress of each activity;
- Forum: Within the course, there is an access to the forums in which consultations can be made, solve doubts, and participate with the other participants;
- Blog: In this section, you access a specific blog of the MOOC course that is administered by the teaching team and in which you can include information about the development of the course or information of interest or to expand knowledge.

At the end of each module, an evaluation system will be carried out and at the end of the course; in some of the courses, there will be a final work to be done. In this sense, and in relation to the certifications, MiriadaX offers two types of different certifications:

- Certificate of participation: Achieved when the student exceeds at least 75% of the modules of the course. Recognize your participation in the

MOOC and you can download it as a diploma in PDF format and as a badge;

- Certificate of improvement: You can get this certificate; after payment of your fee (€ 40), those registered persons who request it and who exceed 100% of course modules. It can be downloaded as a diploma in PDF format and as a badge;

The platform has the possibility of accessing through mobile devices using an app for both Android and iOS.

6.5 MOOC Platforms: Comparative Analysis

Then, in Table 6.1, as a conclusion and summary, the information commented on in the previous sections is presented regarding the four MOOC platforms analyzed. This facilitates the possibility of comparing the information discussed.

On the other hand, MOOCs are a relatively recent phenomenon (Graham and Fredenberg, 2015). In 2008, the term "MOOC" was introduced in Canada by Dave Cormier who coined the acronym to designate an online course conducted by George Siemens and Stephen Downes. The course entitled "Connectivism and Connective Knowledge" was performed by 25 students who paid their tuition and obtained their title, but was followed by free and without accreditation courses followed by 2300 students and the general public via the Internet (Downes, 2012 b, Daniel, 2012). After this experience, the second successful attempt to export this idea materialized in the summer of 2011 in which the Stanford University offered an online "Artificial Intelligence" course in which 58,000 students enrolled. One of the people involved in the project was Sebastian Thrun, later founder of the MOOC platform "Udacity" (https://www.udacity.com) that provides support to universities for the development of open training (Meyer, 2012). The Massachusetts Institute of Technology originally created MITx for the design of this type of courses, but has evolved into a joint platform of Harvard University, UC Berkley, and MIT itself, with the name of EDx (https://www.edx.org).Coursera (https://www.coursera.org) (Lewin, 2012; DeSantis, 2012) which is the platform that has most developed these initiatives and what is being signaled out as the standard-bearer in pedagogical design. Figure 6.5 describes their evolution.

The New York Times called 2012 "The Year of the MOOC" publishing an article that highlighted the great impact of MOOCs and stated that these would become a *tsunami* that would sweep traditional universities (Figure 6.6) (Pappano, 2012).

Table 6.1 Summary of the analyzed MOOC platforms (Coursera, edX, Udacity and MiriadaX)

	Coursera	edX	Udacity	Miriadax
Birth	2011	2012	2012	2013
Founders	Stanford University	MIT and Harvard University	Private teaching initiative	Telefónica and Banco Santander
Members/partners	Universities fundamentally (148)	Universities fundamentally	Multinationals of the technology sector	Universities fundamentally
Multi-language	YES	YES	YES	NO (only Spanish and Portuguese)
Includes Castilian	YES	YES	DO NOT	YES
Themes treated	Multiple (10)	Multiple (30)	Fundamentally technology (7)	Multiple (27)
Price of courses	Free and paid	Gratuitous	Free and paid	Gratuitous
Certification	Previous payment	Previous payment	Previous payment	Previous payment in the case of the certificate of improvement, but with option to free certification in the certificate of participation.
The level of difficulty of the MOOC courses is indicated	Do not	Yes	Yes	Do not
Filtering catalog	Basic	Advanced	Advanced	Very basic
Interface courses	Good	Good	Limited	Good
General information about the course	YES	YES	DO NOT	YES

Navegability structure course	Good	Very good	Poor	Good
Forums	Yes	Yes	Yes	Yes
Type evaluation	Questionnaires and tasks (essay, revised, with honors, of programming and mathematics).	Questionnaires and essay assignments	Questionnaires, portfolios and essay tasks	Questionnaires and essay assignments
Section qualifications obtained	No, within each activity	It has a progress section where information about the evaluation is displayed	No, within each activity	It has a specific section on qualifications and is also shown in
Blog course	Do not	Do not	Do not	Yes
Download resources	Do not	Do not	Yes	Do not
Support	Not highlighted	Yes, featured	Do not	Yes, featured
App for mobile devices	iOS and Android	iOS and Android	iOS and Android	iOS and Android

Source: Martín- Padilla and Ramírez- Fernández (2016).

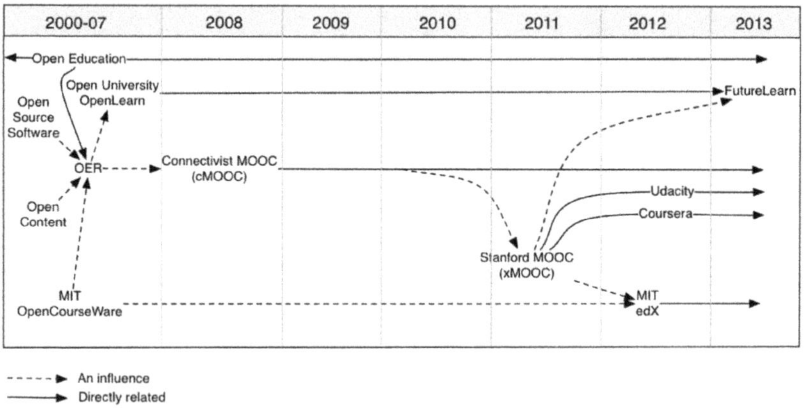

Figure 6.5 Timeline of the genesis of MOOCs and open formation. (White Paper Fountain "MOOC and Open Education: Implications for Higher Education").

In the Horizon report, led by the New Media Consortium and Educause, there is a statement that provides a prospective study of the use of technologies and educational trends in the future of different countries. In its ninth edition (Johnson et al., 2013), the incidence of MOOCs in the current educational panorama stands out. Likewise, the Ibero-American edition oriented to higher education, joint initiative of the "eLearn Center" of the UOC and the New Media Consortium, indicates that the "massive open courses" will be implemented in our institutions of Higher Education in a horizon of four to five years (Durall et al., 2012).

6.6 Additional Resources About MOOCs

With the growth of MOOCs worldwide, in addition to platforms such as those analyzed, other web services that offer information related to this type of courses have also appeared. Among them are localization services and search of courses between platforms, as well as others that serve to list the accreditation obtained by their users in regulated education and other types of training.

Khan Academy[1]

The Khan Academy is a nonprofit educational organization created in 2006, by Salman Khan (professor, computer scientist, and American electrical

[1]Khan Academy. https://www.khanacademy.org/

Figure 6.6 The New York Times (2012 MOOC Year).
Source: http://nyti.ms/ShTBdq

engineer). With the mission of "providing high quality education to anyone, anywhere", the website provides a free online collection in the form of micro classes in video tutorials format stored on YouTube, with different themes: *mathematics, history, finance, physics, chemistry, biology, astronomy, and economics.*

It has also incorporated a summary of student progress that allows teachers to deepen the profile of a person and find out what issues are becoming problematic. The students can make use of an extensive library of contents, including interactive challenges, evaluations, and videos, from any device with access to the network. Teachers and families can easily observe everything the students are learning at the Khan Academy.

It also allows students to monitor their evaluation on the platform and in the learning modules. The personal profile allows you to have a panoramic view of each exercise and problem on which you have worked. It has

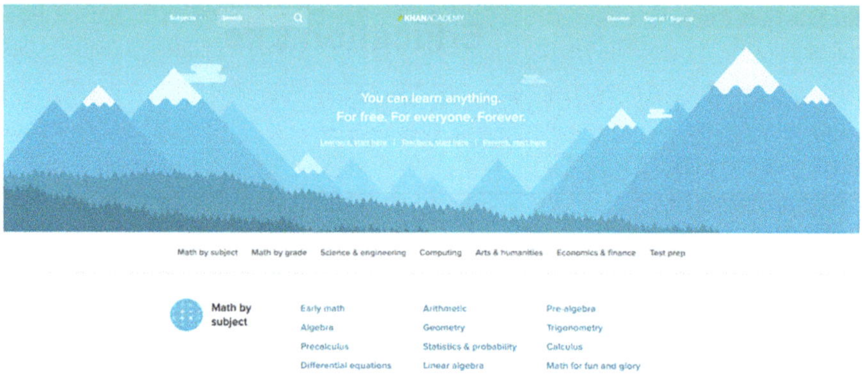

Figure 6.7 Home page of the Khan Academy Platform.
Source: https://www.khanacademy.org/

arbitrated a "crowd-funding" system that allows the person to contribute to the platform through different options: online donation, send a check by mail, make an inventory gift, bank transfer, or donate Bitcoins.

My Education Path[2]

One of the aggregators of courses that allow you to search for courses in some of the main MOOC platforms. My Education Path defines its mission as helping to find free alternatives to high-cost university courses.

In addition to this function, My Education Path offers the possibility of searching exam centers that certify knowledge through the MOOC courses.

Class Central[3]

On the home page, Class Central displays a text box where you can perform a course search. There is also a list of the courses that will start next, showing the name of the course, the name of the instructor, the area to which the course belongs, the starting date, its duration, and the name of the platform that offers it.

Currently, Class Central shows courses from the main American MOOC platforms.

[2]My Education Path. http://myeducationpath.com/courses/

[3]Class Central. http://www.class-central.com

No Excuse List[4]

With a simple design, "No Excuse List" allows locating courses hosted in another group of platforms. To see the full directory of courses, simply follow the link by clicking on the word "here".

In this directory appear many of the best-known educational platforms organized according to the educational environment to which they dedicate their main activity: "Academics", "Art", "Computer Programming", "Languages", "Music", etc.

Tutellus[5]

It is a Spanish collaborative platform that adds more than 4,000 video courses and MOOCs from different Universities, Business Schools and users in Spain and Latin America. We can attend any of the recommended MOOCs or give video courses or classes in our city, always having access to an infinity of free courses when registered on the platform.

6.7 Search Engines for MOOC Courses

In addition to the resources provided, specialized search engines are available in MOOC courses to which interested persons can go to locate courses related to the topics personal interest. Below is a list of the most outstanding thematic search engines (Figures 6.8, 6.9, and 6.10):

MOOC.es[6]
Open Education Europe[7]
Course Buffet[8]

6.8 International Resources on MOOC

The search engine and European web page "Open Education Europe" offers a series of very interesting resources that complement the offer presented here, which we present below.

---

[4] No Excuse List. http://noexcuselist.com
[5] Tutellus. http://www.tutellus.com/aprende
[6] MOOC.es. http://www.mooc.es/
[7] Open Education Europe. http://bit.ly/2oyYwyF
[8] Course Buffet. http://www.coursebuffet.com/

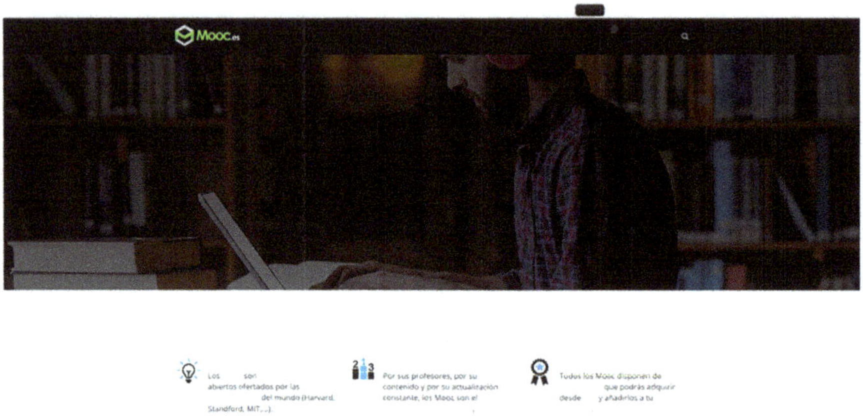

Figure 6.8 Home page of the MOOC.es search engine.

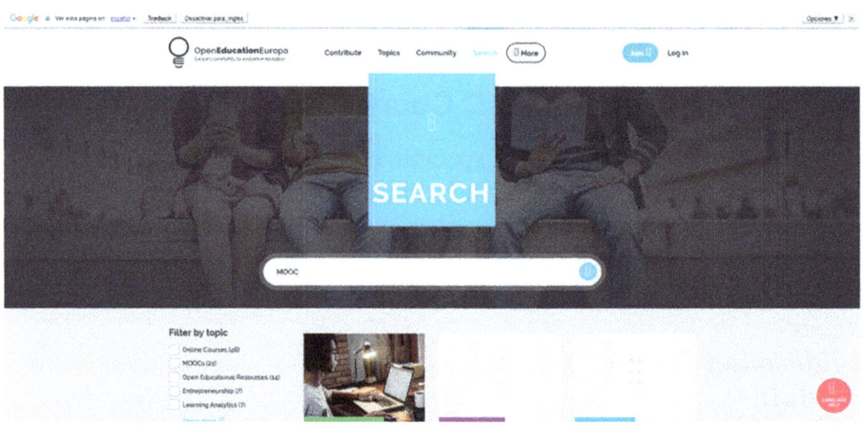

Figure 6.9 Home page of the Open Education Europe search engine.

ALISON[9]

ALISON (Advance Learning Interactive Systems Online) was founded in Galway (Ireland) by Mike Freerick. It is a platform that has more than 600 courses and a community of three million students in more than 200 countries. Since 2007, it has been in operation and more than 350,000 people have already obtained a certificate of training in one of their courses.

---

[9]Alison. http://alison.com/

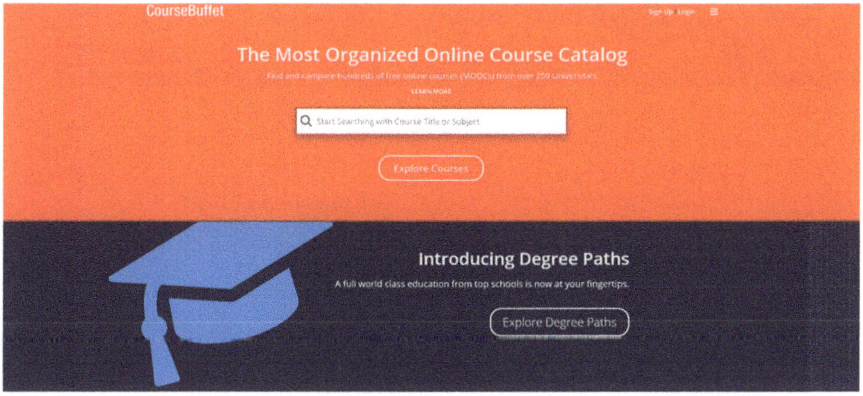

Figure 6.10 Home page of the Course Buffet search engine.

Figure 6.11 Alison home page.

The points that characterize the platform is that its main means of monetization is billing for advertising and visitor traffic. In addition, it offers the possibility for anyone to design a course and become part of this community.

First Business MOOC[10]

This is a platform oriented to the development of courses related to the world of finance and business entrepreneurship.

[10]First Business MOOC. http://firstbusinessmooc.org/

Figure 6.12 Main page of First Business MOOC.

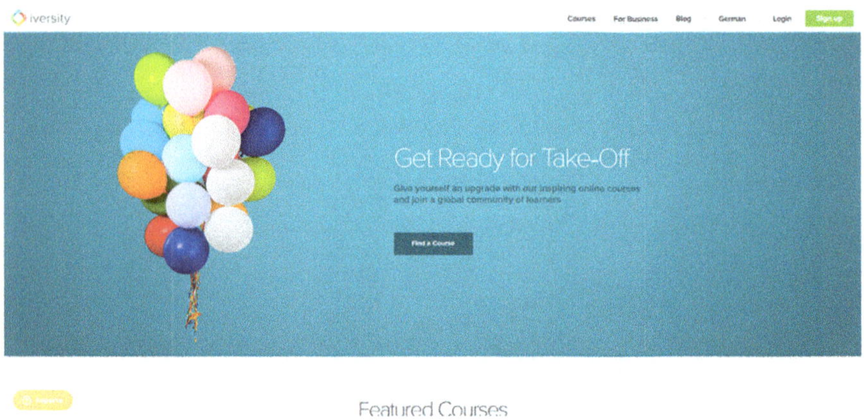

Figure 6.13 Iversity homepage.

Iversity

This is a platform of MOOC courses with a wide range of technology and engineering courses.

Other interesting resources are as follows:

Finally, it is noteworthy that most of the MOOC platforms use social networks to disseminate their tasks, optimize their service strategies, and promote their professional and academic activity (Vázquez-Cano, López-Meneses and Sevillano-García, 2017).

Table 6.2 Other interesting resources about MOOC courses

Resource	Link
CN MOOC MOOC courses in Chinese	http://cnmooc.org/
CN MOOC MOOC courses in Arabic	https://edraak.org/ http://www.rwaq.org/
JMOOC Japanese MOOC courses	http://www.jmooc.jp/
Open Learning Create a MOOC platform	https://www.openlearning.com/
OERu Teaching with open resources	http://oeru.org/

6.9 The Technological Observatories (OT)

In the globalized world in which we find ourselves, Information and Communication Technologies have taken a preponderant role by becoming fundamental tools for the interconnection of societies and in turn by facilitating the flow of knowledge. In this sense, it is necessary that knowledge can be shared, used effectively, and that there are channels for the improvement of knowledge acquisition, both explicit and implicit or tacit (VV.AA, 2014).

Likewise, the large number of information sources and their accelerated growth make it difficult to find the desired information. It is necessary to find a way to identify the intentions and needs of users to reduce the work of searching for information and that it can be obtained more quickly and accurately (Moreno-Espino et al., 2013). Observatories can be an optimal alternative.

The term "observatory" has been found in the recent years, more and more frequently, by scientists, journalists, and politicians from Europe and Latin America. National, regional, and local public administrations and unions and academic institutions and foundations have designed observatories of different types to systematically monitor the progress of a sector or a specific problem. There are observatories related to the most diverse topics: racism and xenophobia, immigration, industrial relations, technology, the environment, or gender violence. Even the authorities of the Louvre Museum have launched their own observatory in order to know in more detail who visits the famous picture gallery (Albornoz and Herschman, 2007).

The dictionary of the RAE (Royal Spanish Academy) defines observatory as a "place or position that serves to make observations".

According to Maiorano (2003), observatories are auxiliary, collegiate, and plural integration organizations that must provide better information to public opinion and encourage decision-making by the responsible authorities. On the other hand, Correa and Castellanos (2014) point out that observatories are reflection spaces based on reality; they allow to align the information and its conservation in specific fields; its indicators and results address situations in context to better understand them and even foresee future effects for the good of the people interested in their object of study or for society.

Also, an observatory is an organism created by a collective in order to follow the evolution of a phenomenon of general interest. From public, national, regional, and local administrations, academic institutions, trade unions, companies, foundations, and civil society organizations have promoted and supported the creation of these bodies that contribute to institutional performance through specific and proactive statements, after studying, recording, and analyzing the situation and evolution of a specific topic (VV.AA, 2014).

According to the United Nations Development Program (UNDP, 2004), the work of an observatory, in general terms, is related to the following areas of work:

- Collection of data and development of databases;
- Methodologies to encode, classify and categorize data;
- Connection of people/organizations working in similar areas;
- Specific applications of the new technical tools;
- Analysis of trends/publications.

The Observatory should initiate a permanent process of dialogue and feedback with those who have the capacity to make decisions, with whom the Observatory's findings can be discussed. In this dialogue, those who manage the Observatory should be open to constructive criticism of the observed, so that both the indicators and the categories of analysis and the techniques used in the processes can be improved (VV.AA, 2011). Then a conceptual approach of Technological Observatory (OT) is made for a better understanding of the object of study of the present doctoral thesis.

A Technological Observatory (OT) is a space in which projects related to the use, implementation, application, and appropriation of Information and Communication Technologies (ICT) are worked on, in order to potentiate technological development in different entities and areas generating economic and social impacts (Torres and Martínez, 2014).

In this sense, De la Vega (2007) indicates that a Technological Observatory (hereinafter OT) generates a knowledge with a high level of importance to be current and novel, which can be used by recipients who have an interest in that information.

A digital observatory is a space in which projects related to the use, implementation, application, and appropriation of ICTs are worked on, in order to promote technological development in entities from different areas, generating economic and social impacts (Torres and Martínez, 2014). In this sense, each OT is a reference in the subject that it processes, giving each sector of the market reliable and important information for users, dynamically, periodically and updated (Bouza et al., 2010). But this information is not personalized; it is for users with common goals and not with specific detailed goals (De la Vega, 2007). That is to say, the Digital Observatories have the mission to organize the information, according to the theme, date, etc., facilitating their access in an effective way.

It can be concluded that the OT measures and processes elements concerning a topic (Pérez-Acosta and Moreno, 2014) and facilitates the work of searching for relevant information that is relevant to the interests of the users, thanks to the integration in a tool that seeks information circumscribed to specific topics, which provides reports, summaries, and alerts that allow users to make decisions (Bouza et al., 2010). And, finally, in its dissemination work, the observatory must deliver the results of its work to be analyzed and used by society (Téllez and Rodríguez, 2014).

Some of the most relevant Technological Observatories are the Technological Observatory of the National Institute of Educational Technologies and Teacher Training[11], or the African Observatory of Science Technology and Innovation[12], or the European Information Technology Observatory[13] (EITO), among others; but the one developed from these by some authors of this book is MOOCservatorio®.

First Business MOOC[14]

It is a platform oriented to the development of courses related to the world of finance and business entrepreneurship.

[11]Technological Observatory of the National Institute of Educational Technologies and Teacher Training. http://recursostic.educacion.es/observatorio/web

[12]African Observatory of Science Technology and Innovation. http://aosti.org

[13]European Information Technology Observatory. http://www.eito.com

[14]First Business MOOC. http://firstbusinessmooc.org/

7

MOOC Reflections on the Future

The MOOCs, as it was proposed in the previous chapters, do not yet know what they are going to end up becoming, in general terms, whether in a permanent formative modality in its multiple possibilities in the university environment, as is already happening in the formative proposals for the teaching staff of basic education by the INTEF; in a micro-offer of credits associated with future accreditations, or in a horizontal movement between experts and non-experts sharing knowledge and certifying among themselves the competencies achieved.

In our opinion, they will survive and specialize, that is, they will keep formats closer to the academy because of their low cost and access to a meritocracy where the monopoly is held by public and private higher education institutions, and the latter will increase. and each time with less demands from the State. But necessarily, the educators will be reorganized, as is already happening with movements of horizontal formation in the basic teachings such as EABE and INNOVADORS; they will be the university professors, who in addition to organizing themselves for research excellence before the criteria of external agencies, will demand a useful and close training from their own colleagues, and who will recognize themselves as experts. And all this in one of the many MOOC formats that are appearing, such as nanomooc, for example; or those that will emerge under the needs and demands of educators.

In addition, it will tend to the supra-specialized networks that will take advantage of the MOOC formats for the reproduction itself, and that will have resources of high academic impact, that is, with journals indexed in the first quartile of JCR.

In general, we project in view of what is currently taking place in the university environment, that the MOOC formats will be a means for different causes, but that they will remain in the panorama of higher education.

Some arguments to be able to affirm such things, are that the technological development of Internet, with the Internet of the Things and the biogadgets, especially the latter, will allow us to have a formative experience totally different from the one that we know at the moment, by access, by interaction, and by assimilation. All this, will first take into account what we think of the development of regulatory regulations and control of access to the NETWORK, as is already happening, regulations that are mainly intended to regulate two unique aspects, the one related to the control of transmedia networks by groups of power, under the most radical postmodern social conditions of the free market, and the most weakened formal democracies. And the reference of homogeneity/equality under the assumptions of human dignity, freedom and protection of the most disadvantaged groups as minors, for example, in order to maintain social control.

Even so, under the strong influence of socioeconomics and politics, MOOCs will be the formats of:

7.1 Homogenization versus Plurality

The MOOCs have enhanced the homogenization of higher education systems, as we saw with platforms such as Coursera, etc. that define a form of transmission and organization of knowledge, and teaching and learning activity. It has endorsed among all the institutions that participate in the platforms that offer mooc courses a single design of teaching, which does not depart from traditional models, if we except the processes of coevaluation, on the other hand, totally logical, if we talk about thousands of students.

On the other hand, the MOOCs have been a starting format, which based on the experience of its users, have been mutated according to their needs, arising a plurality of formats to meet specific demands of training and teaching, for example, the NanoMOOC and the SPOOC.

The first are short courses, between one and twenty hours, and on a very specific content. SPOOCs are courses with no time limit, open to the rhythm of each one as a self-learning guide.

7.2 Individualism versus Differentiation

Traditional MOOCs empower the individual, based on content or task, directing forms of participation "of assigned but informed" the fourth of eight steps, according to Roger Hart's (1993) ladder of social participation and applied to participation in Open teaching formats, where the students are informed and organized in the course tasks, along the milestones marked until the end of the course. This step is considered the first of a real participation, where the eighth step "Participation in actions conceived by the participants themselves and that have been shared with external agents of the educational community", that is, the participants – teachers and students – agree, design, and carry out the course and open it to external agents to their collaboration and contrast during the period of the same. This type of participation, is currently a utopia, in the MOOC training models, but it would be a training model for socio-educational actions that strengthen communities, generating links and synergies on what it is that takes place in spaces of a different nature, and by teams of people from different areas of the community, an example that approaches this idea would be projects such as the "community lab" and the "rural lab".

Returning to the question of participation, the collaborative CMOOCs or MOOCs that have not been extended in the university platforms of MOOC, as we saw in the previous chapters, although they are the only mooc format, with a more real participation at present, that is, in the sixth step "participation in ideas of external agents of educational innovation shared with the professionals of education that are in the course", where the greatest example is found on Stephen Downes' website http://www.downes.ca/index.html.

Therefore, we understand that the MOOCs intensify individualism, because even in the case that group activities are requested, they are actions directed towards the fulfillment of a goal defined by the task, where each one contributes his/her part to resolve it, but that does not allow one to welcome the pluralism that the student group represents, either in the tasks or in their resolution. In addition, from a didactic point of view, the issue raised in the activity is the teacher's view did not permit more views of the matter, even if the students were expert teachers in their areas, such as in master university teachers.

7.3 Individuality versus the Community

Another aspect about individualism in MOOCs is that due to their massive nature, where the sense of identity of the collective is lost, which in traditional courses is at least because of the fact that the students are of same age, but all of this disappears in the MOOC formats, so that we understand that the MOOCs intensify individualism, because even in the case that group activities are requested, they are actions directed to the fulfillment of a goal defined by the task, therefore of symbolic character, but that is not generated among the students, by one side, a culture of participation as an end in itself to achieve knowledge; and on the other hand, they do not conform to a homogenous collective culture that could evince and agree on common needs and desires.

The traditional MOOC formats, due to the reversals of individualism shown above, and as a consequence of those, exclude a large part of the students who massively enroll in a MOOC course. In other words, on the one hand, they do not respond even remotely to the individual particularities of each of the enrolled students, and, on the other hand, the massively enrolled students are strangers to each other, without a collective identity by which they feel connected, and therefore, feeling alone and without accompaniment or complicity. Aspects, which are not addressed in traditional MOOCs, and which would prevent the massive flight of students.

7.4 Abundance vs. Relevance of Training

Traditional MOOCs have flourished in the last five years as if they were an expression of spring, but not all have been relevant from a formative point of view. The MOOC formats have been incorporated into the higher institutions in part by including a new training offer, and on the other, as a fashion or seasonal trend. This has meant an unequal offer of MOOC courses with respect to the training provided, by an unqualified teaching staff in virtual teaching and who followed the procedures of the virtualization secretariat corresponding to each university without assuming them pedagogically, and also because of unattractive contents, or perhaps due to little interest to the target group.

In addition, the teaching design of the traditional MOOCs becomes openly sequenced in weekly content, with no time to appropriate it and integrate it into their practical knowledge. And among them, forming general ideas that are rarely implicated in detailed analysis, where the students do not find the relevance of the same, because the important thing is to move on to the next requirement. Consequently, the selection of content is in the hands of the teacher, as a common reference of the group, making impossible a coherent rhythm among the topics, the contrast, the questioning and the debate of ideas, which would occur if different sources were handled, since the students select the contents in a way suggested by the teacher.

7.5 MOOC Mythification

The capacity of the MOOCs to condescend to the necessary knowledge required on any subject, will make the formative experience based on the study of the current degrees to be reviled, and on the contrary, the MOOCs will be mystified. This tendency has already been worked on through the creation self-oriented teaching courses where the sum of all of them form a master of their own. The modularity of multiple units to form a whole, is what makes the MOCCs add to that, in addition to their vast scale and low cost.

The MOOC is a large-scale training solution, modular, and also adaptive to changes or newly defined needs; let us review these affirmations, in the first place editions of the courses in MOOC format are repeated until they cease to be of interest, and the students are no longer enrolled. If this happens, you can replace a module and add contents that respond to those that are necessary to make it interesting again, and finally facilitating its adaptation to the new demand. For all these reasons, each specialist will develop modules, which will be selected by the administration of the university to mount new titles.

References

Aguaded, J. I. & Medina-Salguero, R. (2015). Criterios de calidad para la valoración y gestión de MOOC. *RIED. Revista Iberoamericana de Educación a Distancia, 18*(2), 119–143.

Aguaded, J. I. (2002). Nuevos escenarios en los contextos educativos: la sociedad postmoderna, del consumo y la comunicación. *Revista científica electrónica Agora digital, 3*, 1–19.

Aguaded, J. I. (2013). La revolución MOOCs, ¿una nueva educación desde el paradigma tecnológico? *Comunicar, 41*, 7–8.

Aguaded, J. I., López-Meneses, E. & Alonso, L. (2010a). Formación del profesorado y software social. Teacher training and social software. *Revista estudios sobre educación,* 18, 97–114.

Aguaded, J. I., López-Meneses, E. & Alonso, L. (2010b). Innovating with Blogs in University Courses: a Qualitative Study. *The New Educational Review*, 22 (3–4), 103–115.

Aguaded, J. I., Pérez, M. A. & Monescillo, M. (2010). Hacia una integración curricular de las TIC en los centros educativos andaluces de Primaria y Secundaria. *Bordón, 62*(4), 7–23.

Aguaded, J. I., Vázquez-Cano, E. & López-Meneses, E. (2016). El impacto bibliométrico del movimiento MOOC en la Comunidad Científica Española. *Educación XX1, 19*(2), 77–104.

Aguaded, J. I., Vázquez-Cano, E. & Sevillano, M. L. (2013). MOOCs, ¿Turbocapitalismo de redes o altruismo educativo? In *SCOPEO INFORME Num. 2: MOOC: Estado de la situación actual, posibilidades, retos y futuro*, 74–90. Salamanca: Universidad de Salamanca. Servicio de Innovación y Producción Digital.

Aguaded, J. I., Muñiz, C & Santos, N. (2011). Educar con medios tecnológicos. Tecnologías telemáticas en la Universidad de Huelva. In *Libro de Actas I Congreso Internacional "Comunicación y Educación: Estrategias de alfabetización mediática"*. Barcelona: Universidad Autónoma de Barcelona.

Albornoz, L. A. & Herschman, M. (2007). Balance de un proceso iberoamericano. Los observatorios de información, comunicación y cultura. *Telos: Cuadernos de comunicación e innovación*, *72*, 47–59.

Almatrafi, O., Johri, A., & Rangwala, H. (2018). Needle in a haystack: Identifying learner posts that require urgent response in MOOC discussion forums. *Computers & Education*, 118, 1–9.

Almutairi, F., & White, S. (2018). How to measure student engagement in the context of blended-MOOC. *Interactive Technology and Smart Education*, *15*(3), 262–278.

Anderson, T. & McGreal, R. (2012). Disruptive Pedagogies and Technologies in Universities. *Education, Technology and Society*, *15*(4), 380–389.

Anderson, T. (2013). *Promise and/or Peril: MOOC and Open and Distance Education*. Commonwealth of Learning. Retrieved of http://bit.ly/2pmHVyh

Angulo-Rasco, F. J. & Bernal-Bravo, C. (2012) ICT as a discurse of salvation. In. Paraskeva & Torres (edts.) *Globalism and Power. Iberian Education and Curriculum Policies* (pp. 107–121). New York Peter Lang.

Area, M. & Pessoa, T. (2012). De lo sólido a lo líquido: las nuevas alfabetizaciones ante los cambios culturales de la Web 2.0. *Comunicar*, *19*(38), 13–20.

Area, M. (2009). *Manual electrónico: Introducción a la Tecnología Educativa*. Universidad de La Laguna. Retrieved of http://bit.ly/2ptPlCg

Area, M. (2012). Enseñar y aprender con TIC: más allá de las viejas pedagogías. *Aprender para educar con tecnología, 2,* 4–7.

Arís, N. & Comas, M. A. (2011). La formación permanente en el contexto del Espacio Europeo de la Formación Permanente. *Revista de Universidad y Sociedad del Conocimiento (RUSC), 8*(2), 5–13.

Armstrong, L. (2014). 2013-The Year of Ups and Downs for the MOOCs. *Changing Higher Education*. Retrieved of http://goo.gl/SqwGWn

Backhaus, K., & Paulsen, T. (2018). Vom Homo Oeconomicus zum Homo Digitalis–Die Veränderung der Informationsasymmetrien durch die Digitalisierung. In *Marketing Weiterdenken* (pp. 105–122). Wiesbaden: Springer.

Baggaley, J. (2011). *Harmonising Global Education: from Genghis Khan to Facebook*. London: Routledge.

Balfour, S. P. (2013). Assessing writing in MOOCs: Automated essay scoring and Calibrated Peer Review. *Research & Practice in Assessment, 8*(1), 40–48.

Ballesteros, C. & López-Meneses, E. (1998). Educación y Nuevas Tecnologías: un diálogo necesario y una realidad evidente. In Cebrián, M. et al. (Coord.), *Creación de materiales para la Innovación Educativa con Nuevas Tecnologías. Edutec'97.* Málaga: I.C.E. de la Universidad de Málaga.

Ballesteros, C., López-Meneses, E. & Torres, L. (2004). *Las plataformas Virtuales: escenarios alternativos para la formación.* In *I Congreso Internacional sobre Educación y Tecnologías de la Información y la Comunicación, Edutec 2004.*

Ballesteros, F. (2003). *Brecha digital: una herida que requiere intervención.* E-business Center PwC & IESE.

Barnes, C. (2013). MOOCs: The Challenges for Academic Librarians. *Australian Academic & Research Libraries, 44*(3), 163–175.

Barroso, J. & Llorente, M. (2007). La alfabetización tecnológica. In Cabero, J. (Coord.), *Tecnología educativa* (pp. 91–104). Madrid: McGraw-Hill.

Bartolomé-Pina, A. R. & Steffens, K. (2015). ¿Son los MOOC una alternativa de aprendizaje? *Comunicar, 44*, 91–99.

Bartolomé-Pina, A. R. (2013). Qué se puede esperar de los MOOC. *Comunicación y Pedagogía, 269–270*, 49–55.

Bauman, Z. (2006). *Modernidad líquida.* Buenos Aires: Fondo de Cultura Económica.

Bouchard, P. (2011). Network promises and their implications. In The impact of social networks on teaching and learning]. *Revista de Universidad y Sociedad del Conocimiento (RUSC), 8*(1), 288–302.

Bouza, O., Gutiérrez-Álvarez, M. & Raposo, R. (2010). Sistematización de la Vigilancia Científica y Tecnológica en organizaciones cubanas. *Ciencias de la Información, 41*(2), 53–57.

Boxall, M. (2012). *MOOC: a massive opportunity for higher education, or digital hype?* Retrieved of http://bit.ly/2pGJ0Et

Burkle, M. (2004). El aprendizaje on-line: oportunidades y retos en instituciones politécnicas. *Comunicar, 37*(XIX), 45–53.

Bustamante, J. (2001). Hacia la cuarta generación de Derechos Humanos: repensando la condición humana en la sociedad tecnológica, *Revista Iberoamericana de Ciencia, Tecnología, Sociedad e Innovación*, 1. Retrieved of http://bit.ly/2qnuZYj

Caballo, M. B., Caride, J. A., Gradaílle, R. & Pose, H. M. (2014). Los Massive Open Online Courses (MOOCS) como extensión universitaria. *Profesorado. Revista de Currículum y Formación del Profesorado, 18*(1), 43–61.

Cabero, J. & Aguaded, J. I. (2003). Presentación: tecnologías en la era de la globalización. *Comunicar,* 11(21), 12–14.

Cabero, J. (1995). Televisión: usos didácticos convencionales. In J. Rodríguez Diéguez y O. Sáenz Barrio (Eds.), *Tecnología Educativa. Nuevas tecnologías aplicadas a la educación,* 213–323. Alcoy: Marfil.

Cabero, J. (2001). *Tecnología educativa. Diseño y utilización de medios en la enseñanza*. Barcelona: Paidós.

Cabero, J. (2003a). La galaxia digital y la educación: los nuevos entornos de aprendizaje. In Aguaded, J. I., *Luces en el laberinto audiovisual,* 102–121. Huelva: Grupo Comunicar.

Cabero, J. (2003b). La utilización de las TICs, nuevos retos para las Universidades. In J. Quesada et al., *I Simposio Iberoamericano de virtualización del aprendizaje y la enseñanza.* San José de Costa Rica: Instituto Tecnológico de Costa Rica.

Cabero, J. (2003c). Las nuevas tecnologías de la información y comunicación como un nuevo espacio para el encuentro entre los pueblos iberoamericanos. *Comunicar, 20*, 159–167.

Cabero, J. (2004). Reflexiones sobre la brecha digital y educación. In F. Soto & J. Rodríguez, *Tecnología, educación y diversidad: retos y realidades de la inclusión digital,* 23–42. Murcia: Consejería de Educación y Cultura. Comunidad Autónoma Región de Murcia.

Cabero, J. (2006a). Bases pedagógicas del e-learning. *Revista Universidad y Sociedad del Conocimiento. (RUSC), 3*, 1–10.

Cabero, J. (2006b). Comunidades virtuales para el aprendizaje. Su utilización en la enseñanza. *EDUTEC, Revista Electrónica de Tecnología Educativa,* 20. Retrieved of http://www.edutec.es/revista/index.php/edutec-e/article/viewFile/510/244

Cabero, J. (2008). La formación en la sociedad del conocimiento. *INDIVISA. Boletín de estudios e investigación, 10*, 13–48.

Cabero, J. (2013). El aprendizaje autorregulado como marco teórico para la aplicación educativa de las comunidades virtuales y los entornos personales de aprendizaje. *Revista Electrónica Teoría de la Educación: Educación y Cultura en la Sociedad de la Información, 14*(2), 133–156.

Cabero, J. (2014). *Los entornos personales de aprendizaje (PLE).* Antequera: IC editorial.

Cabero, J. (2015). Visiones educativas sobre los MOOC. *RIED. Revista Iberoamericana de Educación a Distancia, 18*(2), 39–60.

Cabero, J. (Dir.) et al. (2002). *Las TIC en la Universidad.* Sevilla, MAD.

Cabero, J., López Meneses, E & Llorente, M. C. (2012). E-portafolio universitario como instrumento didáctico 2.0 para la reflexión, evaluación e investigación de la práctica educativa en el espacio europeo de educación superior. *Revista Virtualidad, Educación y Ciencia (VEC), 4*(3), 27–45.

Cabero, J., Llorente, M. C. & Vázquez, A. I. (2014). Las tipologías de MOOC: su diseño e implicaciones educativas. *Profesorado. Revista de Currículum y formación del profesorado, 18*(1), 14–26.

Cabero, J., Marín, V. & Llorente, M. C. (2013). *Desarrollar la competencia digital.* Sevilla: Eduforma.

Calderón-Amador, J. J., Ezeiza, A. & Jimeno-Badiola, M. (2013). La falsa disrupción de los MOOC: La invasión de un modelo obsoleto. In *6º Congreso Internacional de Educación Abierta y Tecnología Ikasnabar´13.* Zalla. Retrieved of http://bit.ly/1MmY9yi

Cañal, P., Ballesteros, C. & López-Meneses, E. (2000). Internet y educación ambiental: una relación controvertida. *Investigación en la Escuela, 41*, 89–101.

Carnoy, M. (2009). Globalización, educación y la Economía del Conocimiento. In Fundación Telefónica, *Globalización y Justicia social. Foro Internacional de Valparaiso 2008,* 157–176. Barcelona: Ariel.

Carrera, J., & Ramírez-Hernández, D. (2018). Innovative Education in MOOC for Sustainability: Learnings and Motivations. *Sustainability, 10*(9), 2990.

Castaño-Garrido, C. & Cabero, J. (2013). *Enseñar y aprender en entornos m-learning*. Madrid: Síntesis.

Castaño-Garrido, C. & Llorente, M. (2007). La organización de los escenarios tecnológicos. La influencia de las TICs en la organización educativa. In Cabero, J. (Coord.), *Tecnología educativa*, 281–296. Madrid: McGraw-Hill.

Castaño-Garrido, C. (2008). *La segunda brecha digital.* Madrid: Cátedra.

Castaño-Garrido, C. (2013). Tendencias en la investigación en MOOC. Primeros resultados. *In 6º Congreso Internacional de Educación Abierta y Tecnología Ikasnabar´13.* Zalla. Retrieved of http://goo.gl/mBKuTi

Castaño-Garrido, C., Maíz-Olazabalaga, I. & Garay Ruiz, U. (2015). Percepción de los participantes sobre el aprendizaje en un MOOC. *RIED. Revista Iberoamericana de Educación a Distancia, 18*(2), 197–221.

Castaño-Muñoz, J., Duart, J. & Teresa, S. (2015). Determinants of Internet use for interactive learning: an exploratory study. *Journal of New Approaches in Educational Research, 4*(1), 25–34.

Castells, M. (1998). *La Era de la Información. Vol. I.* Madrid: Alianza Editorial.

Castells, M. (2000). *La era de la información. Economía, sociedad y cultura. La sociedad en red.* Madrid: Alianza Editorial.

Castells, M. (2001). *La galaxia Internet*, Barcelona: Plaza&Janés.

Castells, M. (2002). *La dimensión cultural de Internet*. Barcelona: Institut de Cultura. Universitat Oberta de Catalunya. Retrieved of http://bit.ly/2oQpW1s

Cataldi, Z. & Cabero, J. (2010) La promoción de competencias en el trabajo grupal con base en tecnologías informáticas y sus implicaciones didácticas. *Píxel-Bit. Revista de Medios y Educación, 37*, 209–224.

Clark, D. (2013). *MOOCs: taxonomy of 8 types of MOOC*. Donald Clark Plan B. Retrieved of http://bit.ly/2qk8svp

Cobos D., Gómez Galán, J., & Sarasola, J. L. (2016). Entre la web social y la disrupción: apuntes para una innovación pedagógica en la Universidad. In D. Cobos, J. Gómez Galán & E. López-Meneses (Eds.). *La Educación Superior en el siglo XXI: nuevas características profesionales y científicas* (pp. 12–34) Cupey: UMET/Innovagogía

Colorado, A., Marín-Díaz, V. & Zavala, Z. (2016). Impacto del grado de apropiación tecnológica en los estudiantes de la Universidad Veracruzana. *International Journal of Educational Research and Innovation (IJERI)*, *5*, 124–137.

Conole, G. (2013). Los MOOC como tecnologías disruptivas: estrategias para mejorar la experiencia de aprendizaje y la calidad de los MOOC. Revista Campus Virtuales. *Revista científica Iberoamericana de Tecnología Educativa*, 2(2), 26–28.

Conole, G. (2015). Designing effective MOOCs. *Educational Media International, 52*(4), 239–252.

Cormier, D. & Siemens, G. (2010). Through the open door: open courses as research, learning, and engagement. *EDUCAUSE Review*, *45*(4), 30–39.

Cormier, D. (2008). *The CCK08 MOOC — Connectivism course, 1/4 way. Dave's Educational Blog*. Retrieved of http://davecormier.com/edblog/2008/10/02/the-cck08-mooc-connectivism-course-14-way/

Correa, G. & Castellanos, L. (2014). Observatorios académicos: hacia una cultura en el uso de la información. *Revista Universidad de La Salle*, *64*, 131–140.

CRUE (1016). *Spanish University in figures 2015–2016*. Madrid: CRUE.

Chamberlin, L. & Lehmann, K. (2011) Twitter in higher education. Cutting-edge. *Technologies in Higher Education, 1*, 375–391.

Chamberlin, L. & Parish, T. (2011). MOOCs: Massive Open Online Courses or Massive and Often Obtuse Courses? *eLearn, 8*. Retrieved of http://bit.ly/2oBjUqn

Chiecher, A. & Lorenzati, K. (2017). Estudiantes y tecnologías. Una visión desde la 'lente' de docentes universitarios. *RIED. Revista Iberoamericana de Educación a Distancia, 20*(1), 261–282.

Christensen, G., Steinmetz, B., Alcorn, B., Bennett, A., Woods, D. & Emanuel, E. J. (2013). *The MOOC phenomenon: who takes Massive Open Online Courses and why?* Retrieved of http://bit.ly/2pqRDlv

Daniel J. (2012). Making Sense of MOOCs: Musings in a Maze of Myth, Paradox and Possibility. *Journal of Interactive Media in Education*, 3. Retrieved of http://doi.org/10.5334/2012-18

Daniel, J., Vázquez-Cano, E. & Gisbert, M. (2015). The future of MOOCs: Adaptative Learning or Business Model? *RUSC. Universities and Knowlwdge Society Journal, 12*(1), 64–73.

Davidson, C. & Goldberg, T. (2009). *The Future of Learning Institutions in a Digital Age*. Chicago: MacArthur Foundation Reports.

De La Torre, A. (2013). *Algunas aportaciones críticas a la moda de los MOOC. educ@contin*. Retrieved of http://bit.ly/2qbupOl

De la Vega, I. (2007). Tipología de Observatorios de Ciencia y Tecnología. Los casos de América Latina y Europa. *Revista Española De Documentación Científica, 30*(4), 545–552.

De Miguel, A. (1979). Universidad, fabrica de parados. *Barcelona, Vicens Vives*.

DeBoer, J., Ho, A., Stump, G. & Breslow, L. (2014). Changing "course:" reconceptualizing educational variables for massive open online courses. *Educational Researcher, 43*(2), 74–84.

Delors, J. (1996). Los cuatro pilares de la educación. In *La educación encierra un tesoro,* 89–103. México: UNESCO.

DeSantis, N. (2012). After leadership crisis fuelled by Distance-Ed Debate, UVa will put free classes online. *Chronicle of Higher Education, 17*. Retrieved of http://chronicle.com/article/After-Leadership-Crisis-Fueled/132917/

DeWaard, I., Abajian, S., Gallagher, M., Hogue, R., Keskin, N., Koutropoulos, A. & Rodríguez, O. (2011). Using mLearning and MOOCs to understand chaos, emergence, and complexity in education. *International Review of Research in Open and Distance Learning, 12*(7), 94–115.

Dillahunt, T., Wang, Z. & Teasley, S. D. (2015). Democratizing Higher Education: Exploring MOOC Use Among Those Who Cannot Afford a Higher Education. *IRROLD, 15*(5), 177–196.

Dillenbourg, P., Fox, A., Kirchner, C., Mitchell, J. & Wirsing, M. (Eds). (2014). *Massive open online courses: current state and perspectives*. Dagstuhl Manifestos. Schloss Dagstuhl — Leibniz Zentrum für Informatik.

Downes, S. (2012a). *The Rise of MOOC*. Retrieved of http://bit.ly/2pq1cPr

Downes, S. (2012b). *Connectivism and Connective Knowledge Essays on meaning and learning Networks*. National Research Council Canada. Retrieved of http://bit.ly/2oTNdk3

Downes, S. (2013). *The Quality of Massive Open Online Courses*. Retrieved of http://www.downes.ca/post/60468

Dron, J. & Ostashewski, N. (2015). Seeking connectivist freedom and instructivist safety in a MOOC. *Educación XX1, 18*(2), 51–76.

Drucker, P. (1994). *Gerencia para el Futuro*. Barcelona: Grupo Editorial Norma.

Drucker, P. (2017). *The age of discontinuity: Guidelines to our changing society*. New York: Routledge.

Duart, J. M., Roig-Vila, R., Mengual, S., & Maseda, M. Á. (2018). La calidad pedagógica de los MOOC a partir de la revisión sistemática de las publicaciones JCR y Scopus (2013–2015). *Revista Española de Pedagogía, 266*, 29–46.

Dublin Descriptors, (2005). *Shared 'Dublin' descriptors for Short Cycle, First Cycle. Second Draft working document*. Dublin: JQI Meeting Dublin 2004.

Durall, E., Gros, B., Maina, M., Johnson, L. & Adams, S. (2012). *Perspectivas tecnológicas: educación superior en Iberoamérica 2012–2017*. Austin-Texas: The New Media Consortium.

Eaton, J. S. (2012). *MOOCs and Accreditation: Focus on the Quality of "Direct-to-Students" Education. Inside Accreditation with the president of Chea*. Retrieved of http://bit.ly/2oQVDIG

Echevarría, J. (2000). La revolución doméstica mete el mundo en casa a través de las nuevas tecnologías. *Consumer, 29*. Retrieved of http://bit.ly/2oQqKDK

Echeverría, J. (1999). *Los Señores del aire: Telépolis y el Tercer Entorno.* Barcelona: Destino.

Echeverría, J. (2010). La Agenda educativa europea y las TIC: 2000–2010. *Revista Española de Educación Comparada, 16*, 75–104.

Egloffstein, M. (2018). Massive Open Online Courses in Digital Workplace Learning. In *Digital Workplace Learning* (pp. 149–166). London: Springer.

Estefanía, J. (2003): *La cara oculta de la prosperidad. Economía para todos.* Madrid: Taurus.

Estévez, J. & García, A. (2015). Las redes sociales para la mejora de la capacidad de emprender y de autoempleo. *International Journal of Educational Research and Innovation (IJERI), 4*, 101–110.

European Commission (2007). *Competencias clave para el aprendizaje permanente. Un marco de referencia europeo.* European Union.

European Commission (2008). *¿Qué es el marco europeo de cualificaciones para el aprendizaje permanente?.* European Union.

European Commission (2016). *Special Eurobarometer 438. E-Communications and the Digital Single Market. Report (May 2016).* European Union.

European Commission (2016). *Special Report 447. Online platforms. Report (June 2016).* European Union.

Fernández-Sierra, J. (1996). La evaluación del profesorado de la Universidad de Almería. Evaluation of teaching staff at the University of Almeria.] Almería: Servicio de Publicaciones.

Ferreres, V. (1997). El desarrollo profesional de los profesores universitarios: la formación permanente, en Rodríguez, J. M. (ed) Seminario sobre formación y evaluación del profesorado universitario, Huelva, ICE de la Universidad de Huelva, 43–71.

Fidalgo, Á., Sein-Echaluce, M. L. & García-Peñalvo, F. J. (2013). MOOC cooperativo. Una integración entre cMOOC y xMOOC. In A. Fidalgo Blanco, y M. L. Sein-Echaluce (eds.), *Actas del II Congreso Internacional sobre Aprendizaje, Innovación y Competitividad, CINAIC 2013* (pp. 481–486). Madrid: Fundación General de la Universidad Politécnica de Madrid.

Finkle, T. A. & Masters, E. (2014). Do MOOC pose a threat to higher education? *Research in Higher Education Journal, 26*, 1–10

Flecha, R. & Elboj, C. (2000). La Educación de Personas Adultas en la sociedad de la información. *Revista Educación XXI. 3*, 141–162.

Flores, R., Ari, F., Inan, F. A. & Arslan-Ari, I. (2012). The Impact of Adapting Content for Students with Individual Differences. *Educational Technology & Society, 15*(3), 251–261.

Floridi, L. (2014). *The fourth revolution: How the infosphere is reshaping human reality*. Oxford: OUP.

Fuente-Cobo, C. (2017). Públicos vulnerables y empoderamiento digital: el reto de una sociedad e-inclusiva. *El profesional de la información, 26*(1), 5–12.

Gallego, D. (2004). La formación del profesorado desde la perspectiva de las organizaciones que aprenden. *Comunicación y Pedagogía. Nuevas Tecnologías y Recursos Didácticos. 195*, 12–19.

García Roca, J., & Mondaza Canal, G. (2002). Jóvenes, universidad y compromiso social una experiencia de inserción comunitaria.

García, F. J., Fidalgo, Á., & Sein, M. L. (2018). An adaptive hybrid MOOC model: Disrupting the MOOC concept in higher education. *Telematics and Informatics*, 35(4), 1018–1030.

García-Ruiz, M. L. (2012). La Universidad Postmoderna y la nueva creación del conocimiento. *Revista Educación XX1. 15*(1) 179–193.

Gee, S. (2012). *MITx — the Fallout Rate*. Retrieved of http://bit.ly/2oQTBsd

Gértrudix, M., Rajas, M. & Álvarez, S. (2017). Metodología de producción para el desarrollo de contenidos audiovisuales y multimedia para MOOC. *RIED. Revista Iberoamericana de Educación a Distancia, 20*(1). Retrieved of http://revistas.uned.es/index.php/ried/article/view/16691/14643

Gisbert-Cervera, M. (2000). El siglo XXI, hacia la sociedad del conocimiento. In J. Cabero Almenara, F. Martínez Sánchez & J. Salinas Ibáñez, *Medios audiovisuales y NNTT para la formación en el S. XXI,* (12–22). Murcia: Edutec.

Gómez Galán, J. & Pérez Parras, J. (2015). Knowledge and Influence of MOOC Courses on Initial Teacher Training. *International Journal of Educational Excellence*, 1(2), 81–99.

Gómez Galán, J. & Pérez Parras, J. (2016c). Initial Teacher Training (ITT) with MOOC: Experiences and Reflections. In E. Corbi, E. López Meneses, M. Tejedor, M. Musello y F. Sirignano (eds.). *Education and Innovation in the University* (pp. 191–205). Nápoles: Suor Orsola Benincasa University Press.

Gómez Galán, J. & Pérez Parras, J. (2016a). El fenómeno MOOC en experiencias de formación inicial del profesorado en españa: enfoque cuantitativo de la problemática. In E. Corbi, M. Tejedor, M. Musello, F. Sirignano & I. McFadden (Eds.). *Nuevos escenarios y perspectivas pedagógicas en el mediterráneo. innovación, nuevas tecnologías y emergencias educativas* (pp. 14–24). Seville: AFOE.

Gómez Galán, J. & Pérez Parras, J. (2016b). Initial Teacher Training (ITT) with MOOC: Experiences and Reflections. In E. Corbi, E. López Meneses, M. Tejedor, M. Musello & F. Sirignano (eds.). *Education and Innovation in the University* (pp. 191–205). Nápoles: Suor Orsola Benincasa University Press.

Gómez Galán, J. & Pérez Parras, J. (2017). Luces y sombras del fenómeno mooc: ¿representan una auténtica innovación educativa? *Revista de Pedagogía*, *36*(102), 237–259.

Gómez Galán, J. (1999). *Tecnologías de la información y la comunicación en el aula.* Madrid: Seamer.

Gómez Galán, J. (2003). *Educar en nuevas tecnologías y medios de comunicación.* Sevilla-Badajoz: Fondo Educación CRE.

Gómez Galán, J. (2009). Tecnología digital para la educación en la sociedad del EEES. In J. I. Aguaded y G. Domínguez (coords.). *La Universidad y las tecnologías de la información y el conocimiento. reflexiones y experiencias.* (pp. 12–41). Sevilla: Editorial Mergablum.

Gómez Galán, J. (2014a). El fenómeno MOOC y la universalidad de la cultura: las nuevas fronteras de la educación superior. *Revista de Curriculum y Formación del Profesorado*, *18*(1), 73–91.

Gómez Galán, J. (2014b). Transformación de la Educación y la Universidad en el Postmodernismo Digital: Nuevos conceptos formativos y científicos. In Durán, F. (coord.). *La Era de las TIC en la nueva docencia* (pp. 171–182). Madrid: McGraw-Hill.

Gómez Galán, J. (2015). Media Education as Theoretical and Practical Paradigm for Digital Literacy: An Interdisciplinary Analysis. *European Journal of Science and Theology*, *11*(3), 31–44 (ISSN: 1842–8517).

Gómez Galán, J. (2016a). Methods in Educational Technology Research. In J. Gomez Galan (Ed.). *Educational Research in Higher Education: Methods and Experiences* (pp. 109–124). Aalborg: River Publishers.

Gómez Galán, J. (2016b). El desafío digital de la Universidad en la Globalización. In J. C. Martínez Coll (Ed.). *Transformación e innovación en las organizaciones* (pp. 28–32). Malaga: Servicios Académicos Intercontinentales S.L.

Gómez Galán, J. (2017a). Educational Research and Teaching Strategies in the Digital Society: A Critical View. In E. López Meneses,

F. Sirignano, M. Reyes, M. Cunzio & J. Gómez Galán. *European Innovations in Education: Research Models and Teaching Applications* (pp. 105–119).

Gómez Galán, J. (2017b). Interacciones Moodle-MOOC: presente y futuro de los modelos de e-learning y b-learning en los contextos universitarios, *Eccos Revista Científica*, *44*, 17–31.

Gómez Galán, J. (2018). Nuevos fenómenos educativos como objeto de investigación científica: de la mochila digital a los cursos MOOC. In O. Ponce, N. Pagán-Maldonado & J. Gómez Galán, J. (2018). Issues de investigación educativa en una era global: nuevas fronteras (pp. 181–198). San Juan: Publicaciones Puertorriqueñas Inc.

Gómez Galán, J. (ed.). (2016c). *Educational Research in Higher Education: Methods and Experiences*. Aalborg: River Publishers.

Gómez Galán, J., López Meneses, E. & Cobos, D. (Eds.) (2016). *La Educación Superior en el siglo xxi: el reto de las TIC*. Cupey: UMET/ Innovagogía.

Gómez, M., Roses, S. & Farias, P. (2012). El uso académico de las redes sociales en universitarios. *Comunicar, 38*, 131–138.

Graham, L. & Fredenberg, V. (2015). Impact of an open online course on the connectivist behaviours of Alaska teachers. *Australasian Journal of Educational Technology*, *31*(2), 140–149.

Guárdia, L., Maina, M. & Sangrá, A. (2013). MOOC Design Principles. A Pedagogical Approach from the Learner's Perspective. *eLearning Papers, 33*. Retrieved of http://bit.ly/2pqZpfb

Guo, P. J., Kim, J. & Rubin, R. (2014). How video production affects student engagement: An empirical study of MOOC videos. *In Proceedings of the first ACM conference on Learning @ scale conference* (L@S '14). ACM, New York, NY, USA, 41–50.

Guzmán, M. D., Correa, R. I., Duarte, A. & Pavón, I. (2004). Profesores en Red. Un estudio sobre los procesos formativos del profesorado de la Universidad de Huelva. In *Edutec'2004. Congreso internacional sobre educación y tecnologías de la información y la comunicación: Educar con tecnologías, de lo excepcional a lo cotidiano. Educar con tecnologías, de lo excepcional a lo cotidiano*. Retrieved of http://www.lmi.ub.es/edutec2004/pdf/164.pdf

GWI Social (2017). *GlobalWebIndex's quarterly report on the latest trends in social networking*. London: Global Web Index

GWI Social (2017). *Trends 2017. The trends to watch in 2017*. London: Global Web Index

Hajnal, P. I. (2018). *Civil society in the information age*. New York: Routledge.
Hart, R. (1993). La participación de los niños. De la participación simbólica a la participación auténtica. Ensayos Innocenti No. 4. Unicef.
Hayes, D. (2017). *Beyond McDonaldization*. New York: Routledge.
Hayes, D., Wynyard, R., and Mandal, L. (2002). *The McDonaldization of Higher Education*. Los Angeles: Praeger.
Herrera, L. (2003). La educación en la era de la información. *In Simposio Internacional de Computación en Educación. SOMECE 2003*. Aguascalientes: México.
Hill, P. (2012). Online Educational Delivery Models: A Descriptive View. *Educause Review*, *47*(6), 85–97.
Hinojo, F. J. (2006). Leadership and Superior Education Educative Space. *The International Journal of Learning, 12*, 147–154.
Hollands, F. & Tirthali, D. (2014). *MOOCs: Expectations and Reality. Full Report*. New York: Columbia University. Retrieved of http://cbcse.org/wordpress/wp-content/uploads/2014/05/MOOCs_Expectations_and_Reality.pdf
Hoxby, C. M. (2014). *The economics of online postsecondary education: MOOCs, nonselective education, and highly selective education*. NBER Working Paper 19816. Retrieved of http://www.nber.org/papers/w19816
Johnson, L., Adams Becker, S., Cummins, M., Estrada, V., Freeman, A. & Ludgate, H. (2013). *NMC Horizon Report: 2013 Higher Education Edition*. Austin-Texas: The New Media Consortium.
Jonasson, J. T. (1999). Traditional University Responds to Society? *Lifelong Learning in Europe*, *4*(4), 235–243.
Jordan, K. (2014). Initial Trends in Enrolment and Completion of Massive Open Online Courses. *The International Review of Research in Open and Distance Learning*, *15*(1), 133–160.
Jung, Y., & Lee, J. (2018). Learning Engagement and Persistence in Massive Open Online Courses (MOOCS). *Computers & Education*, *122*, 9–22.
Karsenti, T. (2013). MOOC: Révolution ou simple effet de mode?/The MOOC: Revolution or just a fad? *International Journal of Technologies in Higher Education, 10*(2), 6–37. DOI:10.7202/1035519ar
Kierkegaard, S. (2010). Twitter thou doeth?. *Computer Law & Security Review, 26*(6), 577–594.
Kop, R., Fournier, H. & Mak, J. S. F. (2011). A pedagogy of abundance or a pedagogy to support human beings? Participant support on Massive

Open Online Courses. *The International Review of Research in Open and Distance Learning, 12*(7), 74–93.

Kruger, K. (2000). Proceso de innovación y difusión del conocimiento en empresas. *Scripta Nova. Revista Electrónica de Geografía y Ciencias Sociales, 69*(31), 1–15.

Lane, J. & Kinser, K. (2012). MOOC's and the McDonaldization of global higher education. *The Chronicle of Higher Education, September 28.* Retrieved of http://bit.ly/2oG7n4X

León Urrutia, M., Vázquez-Cano, E., & López Meneses, E. (2017). Análitica de aprendizaje en MOOC mediante métricas dinámicas en tiempo real, @TIC, 18 doi: 10.7203/attic.18.9927

Lewin, T. (2012). Education Site Expands Slate of Universities and Courses. *New York Times*, September 19. Retrieved of http://nyti.ms/2qnKZd1

Little, G. (2013). Massively Open? *The Journal of Academic Librarianship, 39*(3), 308–309.

Liyanagunawardena, T. R., Adams, A. A. & Williams, S. A. (2013). MOOCs: A systematic study of the published literature 2008–2012. *International Review of Research in Open and Distance Learning, 14*(3), 202–227.

López Meneses, E., & Vázquez-Cano, E. (2017). Los MOOC y su incidencia en el Espacio Europeo de Educación Superior: retos y propuestas desde una perspectiva crítica. Aula Magna 2.0. Revistas Científicas de Educación en Red. Retrieved of https://cuedespyd.hypotheses.org/260

López-Meneses, E. & Ballesteros, C. (2000). Nuevos lenguajes y nuevos tiempos: la comunicación multimedia a través de las redes. In Calderón, M. C., Pérez, E. et al. *Educación y Medios de Comunicación Social: Historia y perspectivas,* 339–345. Sevilla: Running Producción.

López-Meneses, E. & Miranda-Velasco, M. J. (2007). Influencia de la tecnología de la información en el rol del profesorado y en los procesos de enseñanza-aprendizaje. *RIED. Revista Iberoamericana de Educación a Distancia, 10*(1), 51–60.

López-Meneses, E. (2012). *Educador Social, Web 2.0 y Actitud 2.0.* Madrid: Editorial Académica Española.

López-Meneses, E. (2017). El fenómeno MOOC y el futuro de la Universidad. *Fronteras de la Ciencia, 1*, 90–97.

López-Meneses, E., Vázquez Cano, E. & Gómez Galán, J. (2014). Los MOOC: La Globalización y la Innovación del Conocimiento Universitario. (pp. 1475–1483). In D. Cobos et al. (eds). *Innovagogía 2014*. AFOE: Sevilla.

López-Meneses, E., Vázquez-Cano, E. & Román-Graván, P. (2015). Analysis and implications of the impact of MOOC movement in the scientific community: JCR and Scopus (2010–2013). *Comunicar, 44*, 73–80. DOI: 10.3916/C44-2015-08

Lugton, M. (2012). What is a MOOC? What are the different types of MOOC? xMOOCs and cMOOCs. *Reflections*, August 23. Retrieved of http://bit.ly/2oGkenF

Mackness, J., Mak, S. F. J. & Williams, R. (2010). The ideals and reality of participating in a MOOC. In L. Dirckinck-Holmfeld, V. Hodgson, C. Jones, M. de Laat, D. McConnell & T. Ryberg, (Eds.), *Proceedings of the 7th International Conference on Networked Learning 2010*, 266–274. Lancaster: Lancaster University.

Maiorano, J. L. (2003). Los observatorios de derechos humanos como instrumento de fortalecimiento de la sociedad civil. *Revista Probidad, 24*, 10–15.

Mañero-Contrera, J. (2016). Estudio de caso de los sMOOC y su pedagogía en el contexto online. *Revista Mediterránea de Comunicación, 7*(2), 1–10.

Marauri, P. M. (2014). La figura de los facilitadores en los Cursos Online Masivos y Abiertos (COMA/MOOC): nuevo rol profesional para los entornos educativos en abierto. *RIED. Revista Iberoamericana de Educación a Distancia, 17*(1), 35–67.

Marcelo, C. (1993). El perfil del profesor universitario y su formación inicial. In *Ponencias y réplicas/III Jornadas nacionales de Didáctica universitaria, Las Palmas de GC del 23 al 26 de septiembre de 1991* (pp. 191–211). Servicio de Publicaciones.

Marquès, P. (2000). *Impacto de la Sociedad de la Información en el mundo educativo*. Retrieved of http://peremarques.pangea.org/impacto.htm

Marquès, P. (2001). Algunas notas sobre el impacto de las TIC en la universidad. *Educar, 28*, 83–98.

Martí, J. (2012). Tipos de MOOCs. *Xarxatic*. Recuperado de http://bit.ly/2qdHglI

Martín, M. A. & López-Meneses, E. (2012). La Sociedad de la Información y la formación del profesorado: E-actividades y aprendizaje colaborativo. *Revista Iberoamericana de Educación a Distancia. 15*(1), 15–35.

Martín, O., González, F. & García, M. A. (2013). Propuesta de evaluación de la calidad de los MOOC a partir de la Guía Afortic. *Campus Virtuales, 2*(1), 124–132.

Martínez de Rituerto, P. (2014). Figura de los facilitadores en los cursos online, masivos y abiertos (COMA/MOOC): nuevo rol profesional para los entornos educativos en abierto. *RIED, Revista Iberoamericana de Educación a distancia, 17*(1), 35–67.

Martínez, A. & Torres, L. (2013). Los Entornos Personales de Aprendizaje (PLE). Del cómo enseñar al cómo aprender. *EDMETIC, Revista de Educación Mediática y TIC, 2*(1), 41–62.

Martínez-Abad, F., Rodríguez-Conde, M. J. & García-Peñalver, F. J. (2014). Evaluación del impacto del término MOOC vs eLearning en la literatura científica y de divulgación. *Profesorado, 18*(1), 1–17.

Martínez-Sánchez, F. (2007). La sociedad de la información. In Cabero, J. (Coord.), *Tecnología educativa.* Madrid: McGraw-Hill, 1–12.

Martín-Padilla, A. H. & Ramírez-Fernández, M. B. (2016). Los MOOC en la Educación Superior. Un análisis comparativo de plataformas. *Revista Educativa Hekademos, 21*, 7–18.

Martín-Padilla, A. H. (2015). La mundialización de la educación a través de los Observatorios sobre MOOC. In Matas, A., Leiva, J. J., Moreno, N., Martín-Padilla, A.H. & López-Meneses, E. (Coords.), *I Seminario Internacional Científico sobre Innovación docente e Investigación Educativa. 2 y 3 de diciembre de 2015 Universidad de Málaga*, 228–242. Sevilla: AFOE.

Martín-Padilla, A. H., Bernal-Bravo, C., Ramírez-Fernández, M. B. & López-Meneses, E. (2016). Propuesta para el diseño de un observatorio de calidad e innovación sobre MOOC. In López-Meneses, E., Cobos, D. & Gómez Galán, J. (Eds.). *La Educación Superior en el Siglo XXI: Una Reflexión desde y para el Profesorado*. Cupey: UMET Press/Innovagogía.

Mazur, E. (2012). The scientific approach to teaching: Research as a basis for course design. In *Opening keynote presentation to ALT-C 2012 conference*.

McAuley, A., Stewart, B., Siemens, G. & Cormier, D. (2010). *Massive Open Online Courses. Digital ways of knowing and learning. The MOOC Model for Digital Practice*. University of Prince Edward Island. Retrieved of http://bit.ly/2oU1MnH

Méndez-García, C. (2013). Diseño e implementación de cursos abiertos masivos en línea (MOOC): Expectativas y consideraciones prácticas. *RED, Revista de Educación a Distancia, 39*, 1–19.

Mengual-Andrés, S. (2011): La importancia percibida por el profesorado y el alumnado sobre la inclusión de la competencia digital en educación Superior (Tesis doctoral). Alicante, Universidad de Alicante.

Mengual-Andrés, S., Roig-Vila, R. & Lloret-Catalá, C. (2015). Validación del cuestionario de evaluación de la calidad de cursos virtuales adaptado a MOOC. *RIED. Revista Iberoamericana de Educación a Distancia, 18*(2), 145–169.

Mengual-Andrés, S., Vázquez-Cano, E. & López-Meneses, E. (2017). La productividad científica sobre MOOC: aproximación bibliométrica 2012–2016 a través de Scopus. *RIED. Revista Iberoamericana de Educación a Distancia, 20*(1), 39–58.

Meyer, R. (2012). What it's like to teach a MOOC (and what the heck's a MOOC?) Retrieved of http://www.theatlantic.com/technology/archive/2012/07/what-itslike-to-teach-a-mooc-andwhat-the-hecks-a-mooc/260000/

Moreno-Espino, M., Carrasco-Bustamante, A., Rosete-Suárez, A. & Dunia, M. (2013). Apoyo a la toma de decisiones en un Observatorio Tecnológico incorporando proactividad. *Revista Ingeniería Industrial, 34*(3), 293–306.

Moser-Mercer, B. (2014). *MOOCs in fragile contexts*. In U. Cress & C. Delgado Kloos (Eds.), *Proceedings of the European MOOC Stakeholder Summit 2014* (pp. 114–121). Lausan: PAU Education.

Moya, M. (2013). La Educación encierra un tesoro: ¿Los MOOCs/COMA integran los Pilares de la Educación en su modelo de aprendizaje online? In *SCOPEO INFORME 2. MOOC: Estado de la situación actual, posibilidades, retos y futuro*, 157–172.

Natividad, G., Mayes, R., Choi, J. I. & Spector, J. M. (2015). Balancing stable educational goals with changing educational technologies: challenges and opportunities. *E-mentor, 1*(58), 89–94.

Oliver, M., Hernández-Leo, D., Daza, V., Martín, C. & Albó, L. (2014). *Cuaderno: MOOC en España*. Madrid: Cátedra Telefónica-UPF. Social Innovation in Education.

Oncu, S. & Cakir, H. (2011). Research in online learning environments: Priorities and methodologies. *Computers & Education, 57*(1), 1098–1108.

ONTSI (2016). *Informe Anual del Sector TIC y de los Contenidos en España 2016.* Madrid: Ministerio de Energía, Turismo y Agenda digital.

ONTSI (2017). *Perfil sociodemográfico de los internautas. Análisis de datos INE 2016.* Madrid: Ministerio de Energía, Turismo y Agenda digital.

Orellana, D. M. (2007). *Incorporación de las Tecnologías de la Información y Comunicación en la Formación Inicial del Profesorado. Estudio de Caso UPNFM, Honduras* (Tesis doctoral). Universidad de Salamanca, España.

Ortega-Carrillo, J. A. (1997). Estudio inicial del nivel de analfabetismo tecnológico-didáctico de los alumnos/as de la licenciatura de Pedagogía. In M. Lorenzo Delgado; F. Salvador Mata & J. A. Ortega Carrillo. (Ed.), *Organización y dirección de instituciones educativas: perspectivas actuales*, 379–405. Granada: Grupo Editorial Universitario.

Oztok, M., Zingaro, D., Brett, C. & Hewitt, J. (2013). Exploring asynchronous and synchronous tool use in online courses. *Computers & Education, 60*(1): 87–94.

Pappano, L. (2012). Year of the MOOC. *New York Times*, 2 de noviembre. Recuperado de http://nyti.ms/2pu3hfd

Peláez, A. F. & Posada, M. (2013). Autonomía en Estudiantes de Posgrado que participan en un MOOC. Caso Universidad Pontificia Bolivariana. In *Scopeo Informe Nº 2. MOOC: Estado de la situación actual, posibilidades, retos y futuro,* 174–193. Salamanca: Universidad de Salamanca-Centro Internacional de Tecnologías Avanzadas.

Pérez Parras, J. & Gómez Galán, J. (2015). Knowledge and Influence of MOOC Courses on Initial Teacher Training. *International Journal of Educational Excellence*, *1*(2), 81–99.

Pérez, F. (2004). Las universidades en la sociedad del conocimiento: la financiación de la enseñanza superior y la investigación. In J. Hernández (Dir), *La Universidad española en cifras,* 43–64. Madrid: Conferencia de Rectores de las Universidades Españolas. Observatorio Universitario.

Pérez-Acosta, A. & Moreno, M. (2014). Un Observatorio Tecnológico con un enfoque de Inteligencia de Negocio. *Ciencias de la Información. 45*(3), 11–18.

Pilli, O., Admiraal, W., & Salli, A. (2018). MOOCs: Innovation or Stagnation? *Turkish Online Journal of Distance Education*, *19*(3), 169–181.

PNUD. (2004). *Experiencias comparativas-PNUD Honduras. Observatorios de Desarrollo Humano.* Panamá: Programa de las Naciones Unidas para el Desarrollo.

Ponce, O. Pagán-Maldonado N. & Gómez Galán J. (2018). *Philosophy of Educational Research in a Global Era: Challenges and Opportunities for Scientific Effectiveness*. San Juan: Publicaciones Puertorriqueñas.

Ponce, O., Pagán-Maldonado, N. & Gómez Galán, J. (2018). *Issues de Investigación Educativa en una Era Global: Nuevas Fronteras*. San Juan: Publicaciones Puertorriqueñas Inc.

Popenici, S. (2014). *MOOCs–A Tsunami of Promises, Popenici. A space for critical analysis in higher education*. Retrieved of http://bit.ly/2p8CtRk

Prendes, M. P. (2007). Internet aplicado a la educación: estrategias didácticas y metodologías. In J. Cabero, (Coord.), *Las nuevas tecnologías aplicadas a la educación*. Madrid: McGrawHill.

Rabanal, N. G. (2017). Cursos MOOC: un enfoque desde la economía. *RIED. Revista Iberoamericana de Educación a Distancia, 20*(1), 145–160. Retrieved of http://revistas.uned.es/index.php/ried/article/view/16664

Radford, A. W., Robles, J., Cataylo, S., Horn, L., Thornton, J. & Whitfield, K. (2014). The employer potential of MOOCs: a survey of human resource professionals' thinking on MOOCs. *RTI International*. Retrieved of http://bit.ly/2oQfpo0

Rajas, M., Puebla-Martínez, B., & Baños, M. (2018). Formatos audiovisuales emergentes para MOOCs: diseño informativo, educativo y publicitario. *El profesional de la información (EPI)*, 27(2), 312–321.

Ramírez-Baldomero, M. & Salmerón, J. L. (2015). Un instrumento para la evaluación y acreditación de la calidad de los MOOCs. [EduTool®: A tool for evaluating and accrediting the quality of MOOCs]. *Educación XX1*, *18*(2), 97–123.

Ramírez-Fernández, M. (2014a). Propuesta de certificación de calidad de la oferta española educativa de cursos MOOC realizada por el Instituto Nacional de Tecnologías Educativas y de Formación del Profesorado. *International Journal of Educational Research and Innovation, 3*, 121–133.

Ramírez-Fernández, M. (2015b) MOOC appraisal: A quality perspective, *RIED. 18*(2), 171–195.

Ramírez-Fernández, M. B. (2014). *Modelo de reglas difuso para el análisis y evaluación de MOOCS con la norma UNE 66181 de calidad de la formación virtual* (Tesis doctoral). Universidad Pablo de Olavide, Facultad Ciencias Sociales.

Ramírez-Fernández, M. B. (2015a). The MECD Quality Certification Proposal of MOOC Courses. *International Journal of Educational Excellence, 1*(2), 111–123.

Ramírez-Fernández, M. B., Salmerón-Silvera, J. L & López-Meneses, E. (2016). El paradigma de la calidad normativa en el diseño de cursos en línea masivos y abiertos. *Revista DIM, 33*, 1–13.

Ramírez-Fernández, M. B., Salmerón-Silvera, J. L. & López-Meneses, E. (2015). Comparativa entre instrumentos de evaluación de calidad de cursos MOOC: ADECUR vs Normas UNE 66181:2012. *RUSC Universities and Knowledge Society Journal, 12*(1), 131–144.

Raposo-Rivas, M., Martínez-Figueira, E. & Sarmiento-Campos, J. A. (2015). Un estudio sobre los componentes pedagógicos de los cursos online masivos. *Comunicar: Media Education Research Journal, 22*(44), 27–35.

Ravenscroft, N. (2011). Connecting communities through food and farming: the serious business of leisure. *Invited keynote paper to the Leisure Studies Association Annual Conference: Leisure in transition: people, policy and places*. Southampton Solent University.

Rees, J. (2013). *MRI 2013: Interview with Jonathan Rees*. Retrieved of http://bit.ly/2oNI100

Rheingold, H. (2013). MOOCs, Hype, and the Precarious State of Higher Ed: Futurist Bryan Alexander. Retrieved of http://bit.ly/2qcFxgw

Roig-Vila, R., Mengual-Andrés, S. & Suárez-Guerrero, C. (2014). Evaluación de la calidad pedagógica de los MOOC. *Revista Profesorado. Currículum y Formación del Profesorado, 18*(1), 27–41.

Roig-Vila, R., Mondéjar, L. & Lorenzo-Lledó, G. (2016). Redes sociales científicas. La Web social al servicio de la investigación. *International Journal of Educational Research and Innovation*, *5*, 171–183.

Román, P. (2002). *El trabajo colaborativo en redes. Análisis de una experiencia en la R.A.C.S.* (Tesis Doctoral) Universidad de Sevilla, Facultad de Ciencias de la Educación.

Ruiz-Bolivar, C. (2015). El MOOC: ¿un modelo alternativo para la educación universitaria? *Revista Apertura*, *7*(2), 1–14.

Ruiz-Palmero, J., Sánchez, J. & Gómez, M. (2013). Entornos personales de aprendizaje: estado de la situación en la Facultad de Ciencias de la Educación de la Universidad de Málaga. *Pixel-Bit. Revista de medios y Educación, 42*, 171–18

Salinas, J. (2004). Innovación docente y uso de las TIC en la enseñanza universitaria. *RUSC. Universities and Knowledge Society Journal*, *1*(1).

Salmerón, H., Rodríguez, S. & Gutiérrez, C. (2010). Metodologías que optimizan la comunicación en entornos de aprendizaje virtual. *Comunicar, 34*(17), 163–171.

Sánchez, M. L. (2014). Diseño y producción de cursos MOOC como estrategia de aprendizaje cooperativo en un ambiente de educación a distancia. *Revista Didáctica, Innovación y Multimedia, 28,* 1–12.

Sánchez, M. M., León, M. & Davis, H. (2015). Desafíos en la creación, desarrollo e implementación de los MOOC: El curso de Web Science en la Universidad de Southampton. *Comunicar, 44*, 37–44.

Sánchez-Vera, M. M. (2010). *Espacios virtuales para la evaluación de aprendizajes basados en herramientas de Web Semántica.* (Tesis doctoral). Universidad de Murcia, España.

Sancho-Vinuesa, T., Oliver, M. & Gisbert, M. (2015). MOOCS en Cataluña: Un instrumento para la innovación en educación superior. *Educación XX1, 18*(2), 125–146.

Sandeen, C. (2013). Assessment's place in the new MOOC world. *Research & Practice in Assessment, 8*(1), 5–12.

Sangrá, A., González-Sanmamed, M. & Anderson, T. (2015). Metaanálisis de la investigación sobre MOOC en el período 2013–2014. *Educación XXI, 18*(2), 21–49.

Sanz, M. A. (1995). A, B, C, de Internet. *Boletín Red Iris,* 32. Retrieved of http://www.rediris.es/difusion/publicaciones/boletin/28/enfoque1.pdf

SCOPEO (2013). *SCOPEO INFORME Nº 2. MOOC: Estado de la situación actual, posibilidades, retos y futuro.* Salamanca: Universidad de Salamanca-Centro Internacional de Tecnologías Avanzadas.

Schulmeister, R. (2012). *As Undercover Student in MOOCs, Keynote "Campus Innovation und Jonferenztagung".* Hamburg: University of Hamburg. Retrieved of http://bit.ly/2p8WNlm

Schworm, S. & Gruber, H. (2012). E-Learning in universities: Supporting help-seeking processes by instructional prompts. *British Journal of Educational Technology, 43*, 272–281.

Serrano, A. & Martínez, E. (2003). *La brecha digital. Mitos y realidades.* Mexicali: Universidad Autónoma de Baja California.

Sevillano, M. L. (2008). *Nuevas Tecnologías en Educación Social.* Madrid: Mc-Graw Hill.

Shank, P. (2012). Four Typical Online Learning Assessment Mistakes. In R. Kelly (Ed.), *Online Classroom. Report Assessing online learning: Strategies, challenges, opportunities,* 4–6. Magna Publications.

Siemens, G. (2005). Connectivism: A Learning Theory of the Digital Age. *International Journal of Instructional Technology and Distance Learning, 2*(1), 3–10.

Siemens, G. (2007). Connectivism: creating a learning ecology in distributed environments, In Hug, Th. (Ed), *Didactics of microlearning. Concepts, discourses and examples,* 53–68. Múnster: Waxmann.

Siemens, G. (2008). *Learning and knowing in networks: Changing roles for educators and designers*. University of Georgia IT Forum.

Siemens, G. (2010). *Teaching in Social and Technological Networks.* Retrieved of https://www.slideshare.net/gsiemens/tcconline

Siemens, G. (2012). MOOCs are really a platform. *Elearnspace*. Retrieved of http://bit.ly/2piU5Ka

Siemens, G. (2013). Massive Open Online Courses: Innovation in Education? In R. McGreal, W. Kinuthia & S. Marshall (Eds.), *Open Educational Resources: Innovation, Research and Practice,* pp. 5–15. Vancouver: Commonwealth of Learning y Athabasca University.

Silvia-Peña, I. (2014) Utilización de MOOC en la formación docente: ventajas, desventajas y peligros *Profesorado. Revista de currículum y formación del profesorado, 18*(1), 155–166.

Sloep, P. (2012). *On two kinds of MOOCs*. Retrieved of http://bit.ly/2qksWEE

Söllner, M., Bitzer, P., Janson, A., & Leimeister, J. M. (2018). Process is king: Evaluating the performance of technology-mediated learning in vocational software training. *Journal of Information Technology*, *33*(3), 233–253.

Statista (2017). *Leading social networks worldwide as of April 2017, ranked by number of active users (in millions)*. Recuperado de http://bit.ly/2pfw9sc

Steinmueller, E. (2002). Las economías basadas en el conocimiento y las tecnologías de la información y la comunicación. *Revista Internacional de Ciencias Sociales, 171*, 1–17.

Stödberg, U. (2012). A research review of e-assessment. *Assessment and Evaluation in Higher Education, 37*(5), 591–604.

Stracke, C. M., Tan, E., Texeira, A. M., Texeira Pinto, M., & Paz, J. (2018). Building a Common Quality Reference Framework for Improving, Assessing and Comparing MOOC Design. Invited Speech at *MOOC-Maker Global Symposium*. Lisbon: CCB.

TC Analysis (2016). *VII Observatorio de Redes Sociales.* Madrid: TC Analysis.

Tedesco, J. C. (2000). Educación y sociedad del conocimiento y de la Información. In *Encuentro Internacional de Educación Media.* Bogotá.

Teixeira, A., Mota, J., García-Cabot, A., García López, E. & De-Marcos, L. (2016). Un nuevo enfoque basado en competencias para la personalización de MOOCs en un entorno móvil colaborativo en red. *RIED. Revista Iberoamericana de Educación a Distancia, 19*(1), 143–160.

Téllez, J. & Rodríguez, M. (2014). Observatorio en Emprendimiento: una postura desde la Facultad de Ciencias Administrativas y Contables de la Universidad de La Salle. *Revista Universidad de La Salle, 64*, 111–130.

Torres, A. R. & Martínez, J. C. (2014). Análisis y propuesta de implementación de un observatorio tic para un conjunto de mipymes de la localidad de Usaquén (Bogotá) en la Universidad de San Buenaventura. *Ingenium, 15*(29), 124–147.

Torres, D. & Gago, D. (2014). Los MOOC y su papel en la creación de comunidades de aprendizaje y participación. *Revista Iberoamericana de Educación a Distancia, 17*(1), 13–34.

Touve, D. (2012). MOOC's Contradictions. *Inside Higher,* 11 de septiembre *Ed.* Retrieved of http://bit.ly/2oBxolW

Tüñez, M. & García, J. (2012). Las redes sociales como entorno docente: Análisis del uso de Facebook en la docencia universitaria. *Píxel-Bit. Revista de Medios y Educación. Publicación, 41*, 77–92.

UIT (2003). *Cumbre Mundial sobre la Sociedad de la Información. Hojas informativas.* Ginebra: Unión Internacional de Telecomunicaciones. Naciones Unidas.

UNESCO (2012). *Las TIC en la Educación. El aprendizaje móvil.* Ginebra: United Nations Educational, Scientific and Cultural Organization.

UNESCO (2013). *Policy guidelines for mobile learning.* Ginebra: United Nations Educational, Scientific and Cultural Organization.

Urcola-Carrera, L. & Azkue-Irigoyen, I. (2016). Experiencia del MOOC Idea Acción. *Revista de Dirección y Administración de Empresas, 23,* 148–162.

Valverde, J. & López-Meneses, E. (2002). Hacia una sociedad en red: recursos telemáticos para la Educación Especial. In López-Meneses, E., Ballesteros, C., Valverde, J. et al., *Retos de la alfabetización tecnológica en un mundo en Red.* Mérida: Junta de Extremadura. Consejería de Educación, Ciencia y Tecnología.

Valverde, J. (2014). MOOC: una visión crítica desde las Ciencias de la Educación. *Profesorado. Revista de Currículum y formación del profesorado*, *18*(1), 93–111.

Valverde, J. (2015). La formación universitaria en Tecnología Educativa: enfoques, perspectivas e innovación. *RELATEC. Revista Latinoamericana de Tecnología Educativa, 14*(1), 11–16.

Vargas, M. R. (2005). Educar en el conocimiento. *Revista de la Educación Superior, 34*(136), 35–48.

Vázquez-Cano, E. & López-Meneses, E. (2014). Los MOOC y la Educación Superior: La Expansión del Conocimiento. Profesorado. Revista de currículum y formación del profesorado, *18*(1), 3–12.

Vázquez-Cano, E. & López-Meneses, E. (2015). La filosofía educativa de los MOOC y la educación universitaria. *RIED. Revista Iberoamericana de Educación a Distancia, 18*(2), 25–37.

Vázquez-Cano, E. & Sevillano-García, M. L. (2011). *Educadores en red.* Madrid: UNED

Vázquez-Cano, E. & Sevillano-García, M. L. (2013). ICT strategies and tools for the improvement of instructional supervision. The Virtual Supervision. *The Turkish Online Journal of Educational Technology, 12*(1), 77–87.

Vázquez-Cano, E. (2013). The Videoarticle: New Reporting Format in Scientific Journals and its Integration in MOOCs. *Comunicar, 41*, 83–91.

Vázquez-Cano, E. (2015). El reto tecnológico para la sostenibilidad de los massive open online course (MOOC). *Panorama, 9*(17), 51–60.

Vázquez-Cano, E., López-Meneses, E. & Barroso-Osuna, J. (2015). *El futuro de los MOOC: Retos de la formación on-line, masiva y abierta*. Madrid: Síntesis.

Vázquez-Cano, E., López-Meneses, E. & Sarasola, J. L. (2013). *La expansión del conocimiento en abierto: Los MOOCs.* Barcelona: Octaedro.

Vázquez-Cano, E., López-Meneses, E. & Sevillano-García, M. L. (2017). La repercusión del movimiento MOOC en las redes sociales. Un estudio computacional y estadístico en Twitter. *Revista Española de Pedagogía, 75*(266), 39–58.

Vázquez-Cano, E., López-Meneses, E., Méndez-Rey, J. M., Suárez-Guerrero, C., Martín-Padilla, A. H., Román-Graván, P., Gómez-Galán, J. & Revuelta-Domínguez, F. I. (2013): *Guía didáctica sobre los MOOC.* Sevilla: AFOE.

Vázquez-Cano, E., Méndez, J. M., Román, P. & López-Meneses, E. (2013). Diseño y desarrollo del modelo pedagógico de la plataforma educativa "Quantum University Project". *Revista Campus Virtuales, 2*(1), 54–63.

Viberg, O., & Grönlund, Á. (2017). Understanding students' learning practices: challenges for design and integration of mobile technology into distance education. Learning, *Media and Technology*, *42*(3), 357–377.

Villa, A. & Poblete, M. (2007). *Aprendizaje Basado en Competencias. Una propuesta para la evaluación de competencias genéricas*. Bilbao: Mensajero.

Villar, L. M. & Cabero, J. (1997). *Desarrollo profesional docente en nuevas tecnologías de la información y comunicación.* Sevilla: Grupo de Investigación Didáctica.

Vizoso Martín, C.M. (2013) ¿Serán los COMA (MOOC), el futuro del e-learning y el punto de inflexión del sistema educativo actual? *Revista Intenciones, 5*, 1–12.

VV.AA. (2011). *Observatorio de Educación de Adultos en América Latina y el Caribe.* UNESCO.

VV.AA. (2014). *La función social de los Observatorios El caso del Observatorio Latinoamericano de la Administración Pública (OLAP).* México: INAP

Wade, M. C. (2012). A Critique of Connectivism as a Learning Theory. In *Cybergogue (blog)*. Retrieved of http://bit.ly/2oX6WiU

Waite, M., Mackness, J., Roberts, G. & Lovegrove, E. (2013). *Liminal participants & skilled orienteers: A case study of learner participation in a MOOC for new lecturers*. JOLT.

Weissmann, J. (2012). There's something very exciting going on here. *The Atlantic*. Retrieved of http://theatln.tc/2piZ9ON

Welman, B. (2004). *The Internet in Every Life: An introduction.* Toronto: NETLAB

Williams, K. M., Stafford, R. E., Corliss, S. B., & Reilly, E. D. (2018). Examining student characteristics, goals, and engagement in Massive Open Online Courses. *Computers & Education*, *126*, 433–442.

Wolton, D. (2004). *La otra mundialización*. Barcelona: Gedisa.

WSIS (2003). *Cumbre Mundial sobre la Sociedad de la Información.* Génova: ITU.

Wukman, A. (2012). *Coursera Battered with Accusations of Plagiarism and High Drop-Out Rates. Online Colleges.*

Yamba-Yugsi, M., & Luján-Mora, S. (2017). Cursos MOOC: factores que disminuyen el abandono en los participantes. *Enfoque UTE*, *8*(1), 1–15.

Young, J. R. (2012). *Inside the Coursera Contract: How an Upstart Company Might Profit from Free Courses. Chronicle of Higher Education.* Retrieved of http://bit.ly/2p8PREY

Young, R. H. (1993). Teoría crítica de la educación y discurso en el aula. Paidós Ibérica:.

Yousef, A. M. F., Chatti, M. A., Schroeder, U., Wosnitza M. & Jakobs, H. (2014). MOOCs – A Review of the State-of-the-Art. In *Proc. CSEDU 2014 conference*, Vol. 3, 9–20. INSTICC, 2014.

Yousef, A. M. F., Chatti, M. A., Wosnitza, M. & Schroeder, U. (2015). Análisis de clúster de perspectivas de participantes en MOOC. *RUSC. Universities and Knowledge Society Journal, 12*(1), 74–91.

Yuan, L. & Powell, S. (2013). *MOOC and Open Education: Implications for Higher Education*, U.K: Cetis. Retrieved of http://publications.cetis.ac.uk/wp-content/uploads/2013/03/MOOCs-and-OpenEducation.pdf

Zabalza, M. Á. (2002). La enseñanza universitaria: el escenario y sus protagonistas (Vol. 1). Narcea Ediciones.

Zancanaro, A. & Domingues, M. J. (2017). Analysis of the scientific literature on Massive Open Online Courses (MOOCs). *RIED. Revista Iberoamericana de Educación a Distancia, 20*(1), 59–80.

Zapata Ros, M. (2013). Analítica de aprendizaje y personalización. *Campus Virtuales. Revista Científica Iberoamericana de Tecnología Educativa, 2*(2), 88–118.

Zhang, Y. (2013). Benefiting from MOOCs. In A. Herrington, V. Couros & V. Irvine (Eds.), *World Conference on Educational Multimedia, Hypermedia and Telecommunications, 2013*, 1372–1377. AACE. Retrieved of http://goo.gl/Q3pXhZ

Zhu, M., Sari, A., & Lee, M. M. (2018). A systematic review of research methods and topics of the empirical MOOC literature (2014–2016). *The Internet and Higher Education, 37*, 31–39.

Index

About the Authors

José Gómez Galán, PhD, (Metropolitan University, Puerto Rico, USA & Catholic University of Avila, Spain) is currently Director of Research Center on International Cooperation in Educational Development (CICIDE), Metropolitan University, Ana G. Méndez University System (AGMUS), Puerto Rico, USA & Catholic University of Avila, Spain. Professor of Theory and History of Education, University of Extremadura, Spain (on special leave). Visiting Researcher and Professor at several international universities: University of Oxford (UK), University of Minnesota (USA), Università degli Studi La Sapienza of Rome (Italy), etc. Doctor (PhD) in Philosophy and Education (extraordinary Doctoral Award), Doctor (PhD) in Geography and History. Director of various research groups in different academic centers on national and international scale. He has received several major awards for teaching and research, and was awarded the National Educational Research Award (Spain). His main area of research today is focused on the integration of MOOC courses in educational backgrounds.

Antonio H. Martín Padilla, PhD. (Pablo de Olavide University, Spain), is Associate Professor in School of Social Work at the Pablo de Olavide University, Spain. He primary research areas have been cognitive learning research with a focus on educational computing, multimedia-based and knowledge-based learning environments, e-learning, and the development of evaluation and assessment methods for the effectiveness of computer-based technologies. Current research activities comprise among other issues the analysis of individual and group problem solving/learning processes and possible support by means of ICT, and analysis of the use of mobile IT in informal learning settings. He has ample practical experience in the design and evaluation of MOOC courses.

César Bernal Bravo, PhD. (University of Almería, Spain & King Juan Carlos University of Madrid, Spain), is Professor of Educational Technology at the University of Almería, Spain, and the King Juan Carlos University of Madrid, Spain. Research fellow at the CARE (Centre for Applied Research in Education). He holds a PhD in Computer Science Education. He has over one hundred publications among journals, books, chapters, and conferences. Professor Bernal's research examines how computer mediation impacts important individual and organizational outcomes, for example, mobile worker productivity, open source community participation, virtual team effectiveness, and MOOC courses efficacy.

Eloy López Meneses, PhD. (Pablo de Olavide University, Spain), is Professor of Educational Technology, Innovation and Change in the Pablo de Olavide University, Spain. He is a National Teaching Prize, and has won a number of national awards for teaching excellence. He has a research background in pedagogy and has active research in educational technology, online learning and blended learning. He has published widely on the impact of blended learning, online learning, mobile technologies and MOOC Courses. He is Director of Research Group Eduinnovagogía (HUM-971).

For Product Safety Concerns and Information please contact our EU representative GPSR@taylorandfrancis.com
Taylor & Francis Verlag GmbH, Kaufingerstraße 24, 80331 München, Germany

www.ingramcontent.com/pod-product-compliance
Ingram Content Group UK Ltd.
Pitfield, Milton Keynes, MK11 3LW, UK
UKHW022211190926
475702UK00003B/307

9788770220620